SpringerBriefs in Statistics

SpringerBriefs present concise summaries of cutting-edge research and practical applications across a wide spectrum of fields. Featuring compact volumes of 50 to 125 pages, the series covers a range of content from professional to academic. Typical topics might include:

- A timely report of state-of-the art analytical techniques
- A bridge between new research results, as published in journal articles, and a contextual literature review
- A snapshot of a hot or emerging topic
- An in-depth case study or clinical example
- A presentation of core concepts that students must understand in order to make independent contributions

SpringerBriefs in Statistics showcase emerging theory, empirical research, and practical application in Statistics from a global author community.

SpringerBriefs are characterized by fast, global electronic dissemination, standard publishing contracts, standardized manuscript preparation and formatting guidelines, and expedited production schedules.

More information about this series at http://www.springer.com/series/8921

Jungwoo Ryoo • Kurt Winkelmann

Editors

Innovative Learning Environments in STEM Higher Education

Opportunities, Challenges, and Looking Forward

 Springer

Editors
Jungwoo Ryoo
Pennsylvania State University
Altoona, PA, USA

Kurt Winkelmann
Valdosta State University
Valdosta, GA, USA

This book is an open access publication.

ISSN 2191-544X   ISSN 2191-5458   (electronic)
SpringerBriefs in Statistics
ISBN 978-3-030-58947-9   ISBN 978-3-030-58948-6   (eBook)
https://doi.org/10.1007/978-3-030-58948-6

This Springer imprint is published by the registered company Springer Nature Switzerland AG
The registered company address is: Gewerbestrasse 11, 6330 Cham, Switzerland

For my parents, Kyung-sik and Jeong-Kang, wife, Hyeseon, and children, Hojin, Eugene, and Youngjin

Jungwoo Ryoo

For my wife Catherine and daughter Mary
Kurt Winkelmann

Foreword

Imagine a world in which innovation is allowed to drive the educational learning environment. Whether it be the ability to reach out around the globe and have virtual access to the experts or to be able to feel like you are immersed into the actual environment, such a future can exist. It is important to keep these possibilities in mind as we continue to explore in-depth the concepts supporting Innovative Learning Environments.

My initial encounter with the authors of this book stemmed from an invitation to be the opening presenter for a workshop on Innovative Learning Environments. As I prepared the charts to present, I pondered on what would be the appropriate message to share with the participants of this workshop, individuals who have spent a lifetime grounded in integrating emerging technologies and innovations into learning spaces. In preparing the presentation, I thought it is best to share my experiences (as an educator, engineer, and STEM motivator) about learning environments that craved for change, that craved for fresh thought and new perspective, and environments that craved for innovation. The best way I could think of to impact and motivate the participants was to allow the voices of those students I had encountered to speak through me.

One story was that of Patrick who enrolled in my Summer Science Academy disinterested in science. His mother was trying to expose him to STEM to spark an interest while also seeking to improve his math and science skills. However, Patrick resisted applying himself. He hated math and science because (in his eyes) it was too abstract, boring, irrelevant, or as he says "stupid." Once I was able to create activities that simply connected science to everything in his life (his favorite foods, his clothes, his tennis shoes, his bicycle, his video games, etc.), Patrick adjusted his perspective and opened his mind to learning.

There are countless other stories such as the following: the stories of students who resisted group projects because of the frustrations from slackers or the inability to see the value of said projects; or the stories of STEM teachers who were struggling with finding engaging and cohesive content and strategies designed to make learning fun; or even my own disappointment in knowing that the same challenges I faced in the engineering learning environment in 1982 were exactly the same as those faced by my daughter in 2017; and countless others who complained or were disinterested because the learning environment was simply too mundane, too disconnected from the real world, and lacked the benefits of modern technology. What I found in all those experiences was that the students/teachers were unable to connect what they learned/taught in class in a way that allowed them to value learning, to experience learning, to apply the knowledge gained to feed innovation. So, imagine how elated I was to be able to interact and engage with the participants, who are the experts and contributors of this book, as they worked so diligently to create this blueprint for the future of Innovative Learning Environments (ILE). Technologies included in that blueprint are personalized and adaptive learning, multimodal learning formats, extended cross reality, and artificial intelligence and machine learning. These technologies are constantly pushing the envelope in their application in learning environments.

We, as educators and scientists, have to remain vigilant. It is not just about how we learn but also about how that learning impacts the educational environment, the work environment, and moreover the world globally. As a retired NASA engineer, I can envision students learning about the solar system and the stars while virtually existing in the world of outer space. I can also imagine them being able to virtually tour space vehicles and learn about the function of the many instruments and controls on the vehicle. I can even anticipate students being virtually present to learn about the sequences and operating procedures involved in preparing a vehicle for flight real time and so many other scenarios that positively impact learning.

With the recent invasion of the COVID 19 virus causing a push for more virtual learning spaces, the utility of Innovative Learning Environments can no longer be questioned or delayed. We are being thrust into this direction of virtual learning and cannot continue educating our students using traditional methods only. I am convinced that Innovative Learning Environments offer the opportunity for learning to be ever-reaching, richer, deeper, more expansive, and can serve to meet the needs of students and teachers in the future.

We already know the impact of learning experiences that are informative, engaging, and fun. We know that such environments are the foundations for creativity, productivity, and growth and allow learners to feel like they are a creative part of the

world that surrounds them. The building blocks for such an environment are what is detailed in this book and will have an impact on STEM learning for years to come.

NASA, STEM   Virginia Cook Tickles
Washington, DC, USA

Acknowledgments

The editors of this book set out to explore the boundless possibilities available through Innovative Learning Environments (ILEs) in STEM Education in 2018. One major goal of ours was to develop a community of like-minded scholars and practitioners along the process. We are closer to that goal as of this writing, and this book is an embodiment of what such a community can accomplish when inspired and nurtured through a project like X-FILEs (sites.psu.edu/xfiles/). We feel so honored and proud to be leading such a talented group of enthusiastic and passionate educators, administrators, and industry leaders.

X-FILEs organizers: from left to right, Lawrence Ragan, Lorraine Ramirez, Kurt Winkelmann, and Jungwoo Ryoo

This work was supported in part by the US National Science Foundation under Exploring the Future of Innovative Learning Environments in STEM Grant numbers 1848609 and 1848612. Any opinions, findings, and conclusions or recommendations expressed in this material are those of the author(s) and do not necessarily reflect the views of the National Science Foundation (NSF).

Publication of this book was funded in part by the Open Access Subvention Fund and the John H. Evans Library at the Florida Institute of Technology.

In addition to the chapter authors, many others contributed to this book project. These unofficial contributors include those who participated as X-FILEs webinar and on-site workshop attendees, reviewers, and proofreaders.

We want to thank each one of the people below for their help with making this book possible.

Personalized and Adaptive Learning Chapter Contributors

Ann Lindbloom, University of Kansas, Lawrence, KS, USA
Dongho Kim, University of Florida, Gainesville, FL, USA

Multimodal Learning Chapter Contributors

Sharon Lynn Chu, Department of Computer and Information Science and Engineering, University of Florida, Gainesville, FL, USA

Cross Reality Chapter Contributors

Douglas North Cook, Assistant Professor of Immersive Media, Chatham University, Pittsburgh, PA, USA
Lindsey Spalding, STEM Leadership Consultant

Artificial Intelligence/Machine Learning Chapter Contributors

Ronald Rusay, Chemistry Department, Diablo Valley College, Pleasant Hill, CA, USA
Joe Russo, Learning Management and Curriculum Architect, Penn Medicine, University of Pennsylvania Health System, Philadelphia, PA, USA

Workshop Attendees

1. Stephanie August, Professor of Computer Science, Loyola Marymount University, and California State University Los Angeles
2. Ziyad Al-Hinai, CEO, INGINE
3. AZ Bashet, Dean, Instructional Support and Online Education, Eastfield College
4. Adam Brogdon, Student, Penn State University
5. Jill Castek, Associate Professor of Teaching, Learning, and Sociocultural Studies, University of Arizona
6. Doug North Cook, Assistant Professor of Immersive Media, Chatham University
7. Debra Duke, Undergraduate Director, Department of Computer Science, Virginia Commonwealth University
8. Kim Eke, Senior Director of Information Technology at University of Pennsylvania Graduate School of Education
9. Cy Fisher, Student, Mifflin County High School
10. Peggy Fisher, Instructor of Information Sciences and Technology, Penn State University
11. Cristi Ford, Vice President, Training at NeighborWorks America
12. Martin Gallagher, Digital Scholarship Lab Manager, Florida Institute of Technology
13. Kathy Jackson, Adjunct Associate Teaching Professor, Penn State University
14. Tom Marcinkowski, Program Chair and Florida Institute of Technology

15. Tom Oh, Professor of Information Sciences and Technologies, Rochester Institute of Technology
16. Lawrence C. Ragan, Education LLC
17. Lorraine Ramirez Villarin, Ph.D. STEM Education, University of North Georgia
18. Pushpa Ramkrishna, Program Director, Division of Undergraduate Education, National Science Foundation
19. Jungwoo Ryoo, Professor of Information Sciences and Technology, Penn State University
20. Andrew Shean, Chief Academic Officer/Senior Vice President, National Education Partners
21. Lindsey Spalding, STEM Leadership Consultant
22. Deborah Taylor, Educational Consultant and Faculty Affiliate, the University of Kansas
23. Virgina Tickles, (Retired from NASA) Marshall Space Flight Center, Huntsville, AL
24. Partice Trocivia, Associate Director, Learning Design, Teaching and Learning Lab, Harvard University, Graduate School of Education
25. Barbara Truman, Strategic Advisor, Immersive Learning and Collaboration, University of Central Florida
26. Kurt Winkelmann, Professor of Chemistry, Valdosta State University
27. Dolly Womack, Executive Director, Strategic Partnerships, Pearson
28. Michelle Yeung, Research and Educational Applications Analyst at Loyola Marymount University

We wish to acknowledge and thank Dr. Lorraine Ramirez Villarin for her help organizing the X-FILEs workshop and the X-FILEs Jam.

Jungwoo Ryoo, Ph.D., Professor of Information Sciences and Technology, Penn State University

Kurt Winkelmann, Ph.D., Professor of Chemistry, Valdosta State University

Contents

Introduction

Jungwoo Ryoo and Kurt Winkelmann

1 X-FILEs Approach

The practice of educating students in college-level science, technology, engineering, and math (STEM) subjects is influenced by many factors, including education research, governmental and school policies, financial considerations, technology limitations, and acceptance of innovations by faculty and students. Working together, stakeholders in STEM higher education must find creative ways to address the increasing need for a diverse US workforce with a strong STEM background (President's Council of Advisors on Science and Technology 2012) and the need for a more STEM-literate general population (National Research Council 2012).

In order to help researchers, developers, educators, and other stakeholders find these creative solutions, we conducted the eXploring the Future of Innovative Learning Environments (X-FILEs) project: a series of agenda-setting, interactive, online activities followed by a face-to-face workshop and a writing initiative which led to this book. Our participants considered the following question:

What are the near-term and longer-term impacts, opportunities, challenges, and future research initiatives related to the development and implementation of innovative learning environments (ILEs) in higher education STEM disciplines?

This project included three components: (1) Interactive, online discussions prior to the workshop introduced participants to various innovative learning environments (ILEs) and solicited feedback from a wide range of stakeholders. These discussions formed the basis of the workshop agenda. (2) Participants at our 2-day,

J. Ryoo (✉)
The Pennsylvania State University, Altoona, PA, USA
e-mail: jryoo@psu.edu

K. Winkelmann
Valdosta State University, Valdosta, GA, USA
e-mail: kwinkelmann@valdosta.edu

© The Author(s) 2021
J. Ryoo, K. Winkelmann (eds.), *Innovative Learning Environments in STEM Higher Education*, SpringerBriefs in Statistics,
https://doi.org/10.1007/978-3-030-58948-6_1

face-to-face workshop engaged in creative activities that helped them envision how ILEs can transform STEM higher education. (3) Project leaders (editors of this book) and invited writers continued to synthesize the ideas, questions, challenges, and solutions proposed during the workshop into this book.

The workshop participants discussed and explored four main ILE categories: personalized and adaptive learning, multimodal learning, cross reality (XR), and artificial intelligence (AI) and machine learning (ML). While a particular technology application may fall within one or more of these four categories listed in Table 1, each category is broader than any single educational innovation. Therefore, the usefulness and relevance of our book will endure beyond the shelf life of any current technology application.

We also solicited the views and ideas of subject matter experts and other stakeholders in STEM higher education, including those from traditional 4-year colleges and universities, online programs, and technical and community colleges.

In order to frame the examination of where and how the ILE category may impact the students' experience, we address the following framing questions (FQ) related to the four ILE categories. In particular, we examined the questions from the context of impact on the various aspects of teaching and learning: content presentation, interactions and communications, learner activities, assessment, and co-curricular activities.

FQ1. What are the *opportunities for gains* in the adoption of the ILE category?

FQ2. What *challenges/barriers* exist that may inhibit the impact of the ILE category on the creation of ILEs by 2026?

FQ3. What *implementation strategies* may be used to advance the ILE category in order to realize the impact by 2026?

FQ4. What *research questions* remain to be addressed in order to optimize the impact of the ILE category?

Below are our working definitions of the four domains of teaching and learning:

Table 1 ILE categories and sample technologies and pedagogies

ILE category	Sample technologies and pedagogies of ILEs
Personalized and adaptive learning	Micro-credentialing (badging), self-regulated learning, individualized learning paths, learning analytics, mastery learning, intelligent tutors, student-centered learning
Multimodal learning formats	Digital storytelling, online and blended learning, flipped classroom, game-based learning (GBL), mobile learning, digital publishing, community engagement
Extended/cross reality (XR)	Virtual reality (VR), augmented reality (AR), mixed reality (MR), virtual worlds (VW)
Artificial intelligence (AI) and machine learning (ML)	Intelligent tutoring systems, stealth assessment, autograders, recommender systems, dashboards, peer feedback platform, dynamic scaffolding, peer-to-peer student communication, Internet of Things (IoT)

Content Presentation and Instruction: The substance or content of a course refers to the information about the subject domain that will be conveyed to the learner. The breadth and depth of the course content are selected to meet the instructional needs of the learner in order to achieve the course objectives.

Interactions and Communications: In the teaching and learning context, interactions may be defined as the exchange of information between class participants. Interactions can be between one to many (teacher to class or student to students), one to one (teacher to individual student or student to student), or group-based (teacher to group or group to group).

Teaching, the act of guiding, illuminating, and explaining the course concepts, facts, and experiences to the learner, is principally a function of interactions and communications. Multiple methods are used to teach course content, including the delivery of lectures, readings, viewings, discussions, and hands-on activities. The media formats used to teach the course content are also varied and may include speech, text, video, audio, and hands-on exercises.

Learner Activities: The student activities are methods used to engage the learner with the course objectives. These engagement methods require the learner to apply, practice, and further seek mastery of the course domain. Methods may include assignments such as practice problems, responding to questions, developing or constructing output that represents an understanding of the course concepts. Methods may also include hands-on activities such as lab exercises and experiments, and field studies.

Assessment: Assessing learners, progress in reaching the defined course objectives can be measured as the material is being delivered (formative evaluation) or at critical milestone markers (summative evaluation). Assessment strategies are viewed as instructional in nature as they may reveal to both the teacher and the student areas of needed improvement towards achieving the expressed learning outcomes.

Co-curricular Activities: The learner interacts with the course materials in many ways, internally to the course offering and externally through a variety of methods such as cultural and civic events, educational programs, internship and job experiences and even student-spirit and athletic events. The student experience consists of a wide range of interactions that may consist of athletic, academic, social, and service dimensions. Providing the student with access to a range of these dimensions is a critical link in the overall quality of the student experience.

For each framing question, we consider how the ILE category impacts the presentation of content, the interactions and communications between and among learners and instructors, and the methods and techniques used to conduct assessment of student progress. These four framing questions, combined with the four ILE categories, create a 4x4x5 3D matrix shown in Table 2.

In the following chapters, we will present our in-depth analysis of each of these combinations.

Table 2 FQ-ILE category matrix

	Personalized and adaptive learning	Multimodal learning formats	Cross reality	Artificial intelligence and machine learning
Opportunities for learning gains	Context-setting framework: • Content presentation • Interactions and communications • Learner activities • Assessment • Co-curricular activities			
Challenges and barriers				
Implementation strategies				
Research questions				

2 X-FILEs Workshop Activities

Drs. Ryoo, Ragan, and Winkelmann hosted three online meetings in the months preceding the 2018 workshop. These introduced the technology categories and solicited feedback for each framing question from participants working in small groups. The online discussions were structured to stimulate broad ideas and visions of the potential for ILEs impacting STEM higher education in 2026. Participants included education administrators, instructional designers, STEM researchers and faculty, and industry experts. Some attended the subsequent workshop, while others could only attend the online meetings. A total of 100 people attended the online meetings.

Organizers prepared a list of potential workshop attendees based on members of their own professional networks, educators previously funded by NSF, online meeting attendees, and others who completed the online registration. Organizers selected a diverse group of attendees that included education researchers and practitioners in all four technology areas (Fig. 1). They adjusted the workshop's agenda and structure based on comments, ideas, and feedback from the online meeting participants.

The 2-day, face-to-face workshop in Melbourne, Florida, provided attendees a chance to collaboratively engage in creative activities that helped them exchange ideas and views on the use of ILEs to transform higher STEM education learning. The workshop commenced with a welcome dinner and two keynote talks. Dr. Virginia Tickles spoke about her professional life as a NASA engineer and her experiences as a strong advocate and promoter of STEM education. In the second talk, NSF DUE Program Officer Dr. Pushpa Ramakrishna described the efforts of NSF to create a vision for the future of STEM education, to which this workshop was contributing.

Day 1 started with a "Lightning Strike" question session as an ice breaker activity. Resembling the World Café Method (Brown 2002), participants were encouraged to move between "question stations" that promoted discussions around the four ILE technology categories under study. Once this data was collected, a set of activities allowed for a deeper look of each technology category.

Dr. Larry Ragan, the workshop moderator, divided attendees into five teams, one for each of the aspects of teaching and learning. Participants in each team explored the framing questions related to personalized and adaptive learning. A team lead and recorder compiled the team's thoughts and transcribed them in a pre-structured

Fig. 1 X-FILEs organizers and participants

Fig. 2 X-FILEs group activities

online document. At the conclusion of the discussion, all attendees reconvened and shared highlights of their discussions with each other. This same activity was repeated in the afternoon session for the second category, multimodal learning (Fig. 2).

Day 2 began with participants creating four narratives, each describing "A Day in the Life" of a higher education student in the year 2026. Stories were developed around one of these four assigned personas: (1) unconventional or alternative learner, (2) rural student attending a university fully remotely, (3) residential student, and (4) career-changing adult learner. These stories with "big ideas" were later shared, captured, and recorded.

To explore the last two technology categories, four posters for XR and four posters featuring AI and ML were created. Each contained one framing question along with a wheel depicting the five aspects of teaching and learning. This gallery walk exercise (Carleton College 2018) encouraged all workshop participants to post ideas, reactions, and considerations within each section of the wheel. Teams then distilled and organized the contents of these idea sheets in order to reveal trends and themes and reduce redundant posts. Using this data, teams organized and structured a response for each of the aspects of teaching and learning. Teams reported results to other participants to foster comments, modifications, and agreement.

This day culminated with a student-led discussion panel in order to gain insights into the student perspective of ILEs in STEM education in 2026. These four student panelists included a current high school junior, two undergraduate students, and a graduate student. Following a series of prepared questions, the workshop participants and panelists engaged in a conversation regarding the expectations and anticipations of the higher education experience in 2026.

Following the event, the workshop organizers invited participants to contribute to a white paper describing the workshop outcomes that later evolved into this book. When necessary, they invited other contributors not present at the workshop to help write and review chapters.

3   ILEs Addressed by X-FILEs Project

Advances in educational technology must be coupled with a deep understanding of how students learn and retain information. Only then can developers create new, innovative ways for students to interact with their course materials that replace or augment traditional approaches inside and outside of the classroom. Although some of the ILEs discussed in this book are heavily dependent on technology, they are also part of sound pedagogical approaches that promote student-centered learning. Some major ILEs listed in Table 1 are briefly described below and may involve a physical space, a virtual space, or a combination of the two.

"Digital storytelling is the modern expression of the ancient art of storytelling" (Barrett 2006) and has become increasingly popular in recent years, as shown well in the examples of YouTube and podcasting. Digital stories are often in the form of multimedia movies that incorporate photos, animations, videos, soundtracks, texts, and narrations. College students are comfortable using diverse platforms and tools (e.g., Unity, OpenSimulator, Xtranormal, or Garry's Mod) to make, play, and share interactive digital stories (Pappas 2013). Many educators are already using digital storytelling in their classrooms to enhance the educational experience of their students. To be able to tell their own digital stories, students have to go through a rigorous process of researching a topic, developing a specific point of view and insights, and finally sharing them in creative ways. This experience is germane to one of the ultimate goals of higher education: producing independent learners.

The Internet of Things (IoT) is a growing field of digital communications that allows objects to communicate with each other, creating smart environments that respond appropriately using data collected by ubiquitous sensors. It is expected that the IoT will have a significant impact on many economic sectors (Manyika et al. 2015). The IoT can impact education in two ways: firstly, students can access more data from interconnected physical objects, either nearby or in a remote location. This data provides students with a richer, more detailed understanding of their environment, and they can use the data to study real-world challenges (i.e., through problem-based learning) (Ali et al. 2017). Secondly, sensors can monitor a student's facial features and other characteristics. This data shows when a student is under stress or not fully engaged in the learning process (Farhan et al. 2018). A teacher could then call for a break or redirect the student's attention back to the activity.

Micro-credentials and digital badges acknowledge accomplishments and skill acquisition at a more granular level than college degrees. Many corporations and online programs already adopt micro-credentialing and badging. IBM offers digital badges for its data science training through its IBM Skills Gateway (IBM 2018). Their digital badges recognize the qualifications of IBM's own employees as well as the public in general. edX, a nonprofit online education consortium of higher education institutions, has introduced MicroMasters programs. Universities are also actively looking into ways to incorporate micro-credentialing and badging into their curriculum (Fain 2016). Some of the benefits include more explicit and organized guidance on students' academic achievements and higher motivation and engagement.

Online learning, in the form of massive open online courses (MOOCs), has existed for many years, but MOOCs remain an innovative method for learning due to their continuous incorporation of new technologies (e.g., wikis, credentialing, and personalized learning algorithms). MOOCs hold great promise for democratizing education, but this ILE has not reached its full potential. Only a fraction of science and engineering subjects are offered by MOOCs. Less than 10% of students enrolled in a MOOC actually finish it (Lakshminarayanan and McBride 2015). Although MOOCs are available to groups who have been traditionally excluded from higher education, most users are employed, male, and live in countries of the developed world (Christensen et al. 2013). In a blended learning environment, the curriculum is presented in both an online format and in a traditional classroom space. For instance, students may complete the "lecture" portion of a chemistry class online but travel to campus to complete their laboratory experiments (Dalgarno et al. 2009).

Flipped classrooms are another example of blended learning. In a flipped classroom setting, students perform technology-rich exercises that prepare them for learner-centered activities during class. Pre-class activities can include videos, interactive computer-based simulations, and social media (Lundin et al. 2018). In-class activities require students to critically think about the subject matter and its applications (O'Flaherty et al. 2015). Flipped classrooms are growing in popularity, and they are used frequently to teach STEM subjects. Student acceptance varies; some enjoy their active role in the classroom, and others resist this format because it requires their greater preparation (McNally et al. 2017).

Game-based learning (GBL) leverages the entertaining experience of computer or noncomputer games to achieve an educational objective such as changing attitudes (Stewart et al. 2013) or improving knowledge in subjects such as math (Castellar et al. 2015) and physics (Hamari et al. 2016). Games may be designed with specific learning objectives in mind, or they can be developed independently (e.g., a commercially available video game with no particular educational goal) that nonetheless can lead to users achieving desired cognitive or affective learning outcomes. Purposefully designed educational games are known as serious games. Their use has been shown to improve student learning (Clark et al. 2016; Vlachopoulos and Makri 2017), and best practices for studying their effectiveness are available. However, more rigorous assessment methods are necessary in order to clearly understand how GBL benefits students (All et al. 2016).

Virtual environments positively impact students' attitudes and learning (Hew and Cheung 2010; Kim and Baylor 2015; All et al. 2016; Winkelmann et al. 2017). Examples of virtual environments relevant to this book include virtual, augmented, and mixed realities and AI-enhanced virtual environments. Game-based learning environments may also have a virtual component.

Computer-based simulations and virtual lab experiments, such as PhET and Late Nite Labs, are effective supplements to many STEM courses (Moore et al. 2014; Pyatt and Sims 2012; Finkelstein et al. 2005; Reece and Butler 2017). Immersive virtual reality (VR) allows the user to experience virtual worlds more deeply by viewing the virtual world through stereoscopic visual displays within a headset. This blocks out other sensory distractions and provides a truly three-dimensional view of the virtual objects and other avatars within the VR program. Motion sensors track users' head and hand motions in real time to create an immersive experience in which users control their avatars and interact with each other in the virtual reality setting. One goal for VR-based education is for students to experience the psychomotor "hands-on" aspect of a real-world lab experience. VR technology is used or is being developed for aviation (Koglbauer et al. 2011; Yin et al. 2014), medicine (Buckley et al. 2012), and training in other fields (Kinateder et al. 2015). Virtual reality will provide students with more opportunities for learning both in and out of the classroom (Ma et al. 2014; Lakshminarayanan and McBride 2015). Augmented reality (AR) allows a user to view and interact with virtual objects layered over real objects when viewed through a headset or mobile device screen. It is a similar technology to VR, and its implementation is generating beneficial outcomes for students (Cai et al. 2014; Borrel and Fourches 2017). In a mixed reality (MR) environment, the virtual content and the real content coexist, and the virtual content responds to changes in the real environment. Collectively, these ILEs are referred to as cross reality (XR). Activities can be collaborative and inquiry-based, depending on how the virtual world is designed.

The use of smartphones and tablets for education is known as mobile learning or m-learning (Lakshminarayanan and McBride 2015). These devices enable students to participate in many other ILEs. For instance, smartphones are useful for collecting and processing IoT data. They provide a display screen for augmented reality devices and run educational game apps. Since mobile devices are now ubiquitous among college students, it is reasonable to expect students to have access to assigned digital media which they view outside of class, as is done in a flipped classroom setting.

Digital publishing refers to the dissemination of digital content, such as text, videos, animations, and interactive visualizations, to the public. Digital publishing has many advantages over its conventional counterpart. For one, it can incorporate diverse digital media into publications. One of the emerging trends in digital publishing is open and independent digital publishing of learning knowledge. For example, version-controlled git-authoring (github.com) and open educational resources (OER) are emerging technologies in digital publishing (OER Commons 2018). Many believe that OERs are the future of educational content due to their more engaging, customizable, and affordable nature.

Intelligent virtual environments that integrate artificial intelligence techniques into the virtual environments may include XR and intelligent tutoring systems. For example, intelligent virtual reality systems (IVRS) incorporate AI algorithms into virtual reality (Aylett and Cavazza 2001). Advances in AI have allowed for the development of virtual environments that incorporate agents, eHealth-related devices, human actors, and emotions, projecting them virtually and managing the interaction between all the elements (Rincon et al. 2017). In these environments, the simulation and detection of human emotions can be used for the improvement of the decision-making processes of the developed entities. Machine learning occurs as the AI system adjusts its own algorithms in order to achieve greater success at its task. AI and ML have tremendous potential to enhance creative inquiry and informal learning, as well as adaptive learning through intelligent tutoring systems. One example is a web-based AI tutoring system in Tunisia that recognizes facial expressions as students progress through science experiments that they can access from anywhere (Adams Becker et al. 2016). Though most of the educational AI software is still in the development stages, advancing technologies could drastically alter the landscape of how students learn (Lynch 2016).

Stealth assessment is the evaluation of a learner's performance in a manner that is unobtrusive, using metrics embedded in the course material (Shute and Kim 2014). Students learn without realizing that they are being constantly monitored for their progress. Another benefit is the immediacy of feedback. The educational content can adapt to how well students are doing at a given moment. Technologies now allow educators to conduct stealth assessment without the concerns of costs or system performance. Educators can closely examine the learning process of individual learners in a group by observing a student's collaborations and contributions without disrupting the group.

4 Education Technology Trends Identified Prior to the X-FILEs Project

A discussion of the future use of ILEs in STEM education brings to mind a cautionary Danish proverb: It is difficult to make predictions, especially about the future. Therefore, it is worthwhile to look back at past predictions in order to see how useful such predictions might be, which predictions were in fact credible, and how intervening events derailed expectations of other technologies.

A ready source of informed predictions is provided by the annual New Media Consortium (NMC)'s Horizon Report for the use of technology in higher education. In addition to identifying and analyzing challenges and trends in higher education (not just STEM education), the Horizon Report describes technologies that are poised for widespread adoption in the near, medium, and long term. The 2013 and 2014 Horizon reports (Johnson et al. 2013, 2014) predicted that 3D printing and wearable technologies in 2013 and quantified self and virtual assistants in 2014 were emerging technologies that will be widely adopted in higher education within 4–5 years (2017–2019, during the time period of the X-FILEs project).

3D printing is the design and fabrication of three-dimensional objects by depositing successive layers of the printing material. A user first creates or downloads a digital file that the printer uses to build the desired object. Ceramics, metals, polymer composites containing metal particles or wood fiber, and even paper are commercially available, but the most common printing materials are plastics (von Übel 2019). 3D printing is the most popular technology used in makerspaces—workshops where students and hobbyists collaborate, learn, and use low- and high-tech tools to create. Consumer-grade printers are inexpensive, and websites allow users to share designs. This technology and makerspaces in general are indeed becoming increasingly common in schools, as reflected by the 2018 NMC Horizon Report which placed widespread adoption of makerspaces within a year (Adams Becker et al. 2018). Students can use 3D printing in many STEM activities for chemistry (Pinger et al. 2019), engineering (Chien et al. 2018), and biology (Gordy et al. 2020). A recent review article (Ford and Minshall 2019) describes the current state of this technology in education. The two most common ways of incorporating 3D printing in STEM education are rapid, inexpensive production of models that improve student learning and teach the skills associated with designing the virtual 3D model (e.g., CAD) and operating the 3D printer.

Wearable technology (wearables) is a category of the Internet of Things (IoT) devices that can incorporate other technologies: smartwatches, exercise tracking devices, AR glasses, VR headsets, smart hearing devices (hearables) that interact with virtual assistants or translate speech, and a growing number of proposed medical sensors that track a wearer's health conditions. Sensors within the wearable transmit data to a mobile computing device or to a remote cloud server for analysis and display for the wearer. This is desirable if the wearer wishes to improve quality of life and increase self-awareness. Real-time, passive, personal data collection is the basis for the quantified self-movement noted in the 2014 NMC Horizon Report.

Wearables highlighted in this report include AR glasses and VR headsets. These technologies have shown the most growth in education during the past 5 years. AR and VR in the consumer entertainment industry increased students' acceptance, generated best practices for improving the student's virtual experience, and popularized development tools for creating educational VR experiences.

Potential wearable applications associated with the quantified self have not advanced as rapidly due to a variety of limitations and challenges. These include the need for a continuous wireless connection to a user's phone and the difficulty of displaying information on small screens (Kemper 2019). Medical sensors must first undergo rigorous study and review by healthcare regulatory bodies such as the US Food and Drug Administration (FDA). Collecting, transmitting, and analyzing a wearer's personal health data raise privacy concerns as well. It is not clear how these wearables would be used in educational settings, although students could use such information to improve their diet and sleep schedule, which could impact their ability to learn (Johnson et al. 2014).

Virtual assistants are now well known, thanks to Hey Google, Apple's Siri, and Amazon's Alexa. A user speaks (or types) to the virtual assistant using conversational language, which interprets the words and context and then gathers relevant information from databases or other apps. Virtual assistants use machine learning to improve their understanding of the user's speech and the context of questions. A particular brand of virtual assistant is designed to seamlessly interact with other technologies developed by the same company. These virtual assistants can answer questions about arithmetic, historical dates, and basic science facts. Some virtual assistants allow third-party development of add-on features that can provide more specialized information. A review of the research literature does not show significant implementation of commercially available virtual assistants in higher STEM education.

Conversational agents, more commonly known as chatbots, are another type of virtual assistants that are more widely used inside and outside of education. A chatbot answers questions or provides information about a specific topic, such as a company's chatbot providing customers with frequently requested services. The most well-known example of this was Goel's use of Jill Watson, a virtual teaching assistant designed to answer students' questions about policies and assignments for his online AI course (McFarland 2016). This application has been employed by others as well (Chopra et al. 2016). Educators can teach a subject by employing a chatbot to conduct conversations with students within a virtual world (Heller et al. 2016). This is especially useful in medical schools to help students learn to converse with patients (Carrard et al. 2020). Virtual tutors can help students learn course content outside of class at their own pace (Coronell et al. 2019).

This brief and admittedly inexhaustive review of previous predictions about education technology use shows that many such predictions are accurate, even when made 5 years in advance. The most widely used technologies have commercial applications as well, such as 3D printing, AR and VR, and virtual assistants. Familiarity with technology can encourage its adoption by faculty and acceptance by students. Other factors that promote the use of new technologies are the ability to lower costs, as in the case of 3D printed models and virtual patients, and provide more personalized learning outside the classroom, such as virtual assistants that save time for faculty and increase convenience for students.

5 Conclusion

This book describes:

- Current and future ILE development opportunities, including the potential for improving academic achievement and noncognitive outcomes (e.g., self-efficacy) among students
- Visions toward solutions to technological, logistical, administrative, and societal challenges inhibiting the realization of the potential of ILEs
- Implementation strategies that advance the potential impact of ILEs
- Research domains requiring additional exploration, analysis, testing, and reporting to maximize the impact of ILEs on STEM higher education by 2026

We hope that our book will help guide education policy-makers, researchers, developers, faculty, and practitioners to make informed decisions about the adoption and use of ILEs.

5.1 Overall Impacts

Throughout our project, we take a holistic view of ILE implementation in higher education by considering not only the benefits of ILEs but also addressing the problems that they create and the challenges to their implementation. For instance, how can educators use technology-based ILEs without exacerbating the digital divide among students, and what policies (either within a school or at the state or federal level) must change in order to promote the adoption of ILEs? By thinking beyond the education research findings about ILEs, we are hoping that this book is creating a solid foundation for developing a greater understanding of how society and educational innovations can influence one another. These considerations lead to more challenging problems, but the solutions envisioned in this book will be more realistic and successful when implemented.

5.2 Limitations

A challenge of integrating many of these ILEs into a curriculum is that the physical space is often set up for a traditional lecture or study area, but many academic activities require group work. This necessitates a rearrangement or remodeling of the classroom. Easily movable chairs, accessible electrical outlets for mobile devices, rearrangeable desk space, and whiteboards available for small groups enable students to engage in learning without the limitations imposed by a classroom design. Students value the functional attributes of a learning space and have definite opinions about where they can effectively study (Beckers 2016). Proponents of new education technology must appreciate how it fits into existing physical structures and campus cultures.

5.3 Future of X-FILEs

The X-FILEs project has been evolving and will continue to adapt to the newly emerging challenges. As we finish our book, the world is in the middle of a global pandemic. COVID-19 caused college faculty to quickly change the way they teach traditionally hands-on STEM activities (e.g., lab experiments) to online instructional settings. This is an example of emergency remote teaching (ERT). Through a survey, online discussions, and interviews, the research team will document, curate, and study the ERT experiences of faculty who used innovative learning environments (ILEs) despite having little to no prior experience with them. Potential ILEs include synchronous and asynchronous teaching by video, use of virtual laboratory activities, and other emerging technologies. Selected faculty participants will have limited ILE teaching experience since this group teaches the majority of undergraduate STEM courses and is most in need of assistance implementing these technologies. It is expected that most participants will share the challenges that they faced; others may have discovered unexpected advantages of using ILEs. They will also convey how their experiences shape their attitudes toward future use of ILEs in STEM education.

This is what our focus is for now, but we are certain that our future projects will continue to take us to many other unexplored areas of the creative use of ILEs in addressing challenges in STEM education.

References

Adams Becker S, Freeman A, Hall CG, Cummins M, Yuhnke B (2016) NMC/CoSN horizon report: K—12 Edition 2016. The New Media Consortium. http://cdn.nmc.org/media/2016-nmc-cosn-horizon-report-k12-EN.pdf. Accessed 10 July 2018

Adams Becker S et al (2018) NMC Horizon Report: 2018 Higher Education. EDUCAUSE

Ali M, Bilal HSM, Razzaq MA, Khan J, Lee S, Idris M, Aazam M, Choi T, Han SC, Kang BH (2017) IoTFLiP: IoT-based flipped learning platform for medical education. Digit Commun Netw 3(3):188–194

All A, Castellar EPN, Van Looy J (2016) Assessing the effectiveness of digital game-based learning: best practices. Comput Educ 92:90–103

Aylett R, Cavazza M (2001) Intelligent virtual environments—a state-of-the-art report. In: Eurographics conference, Manchester, UK. https://pdfs.semanticscholar.org/4551/0efe30a1ec 21cdb290e1ace74d3496d4c7e3.pdf. Accessed 22 July 2018

Barrett H (2006) Researching and evaluating digital storytelling as a deep learning tool. In: Crawford C, Carlsen R, McFerrin K, Price J, Weber R, Willis D (eds) Proc. SITE 2006— Society for Information Technology & Teacher Education International Conference. Orlando, Florida, USA: Association for the Advancement of Computing in Education (AACE), pp 647– 654. Retrieved July 19, 2018 from https://www.learntechlib.org/primary/p/22117/

Beckers R (2016) Higher education learning space design: form follows function? In: 15th EuroFM research symposium. http://orbit.dtu.dk/files/124939454/EFMC2016_proceeding. pdf. Accessed 20 July 2018

Borrel A, Fourches D (2017) RealityConvert: a tool for preparing 3D models of biochemical structures for augmented and virtual reality. Bioinformatics 33:3816–3818

Brown J (2002) The world Café: a resource guide for hosting conversations that matter. Whole Systems Associates

Buckley CE, Nugent E, Ryan D, Neary PC (2012) Virtual reality—a new era in surgical training. Virtual reality in psychological, medical and pedagogical applications. In: Eichenberg C (ed), IntechOpen. https://doi.org/10.5772/46415. https://www.intechopen.com/books/virtual-reality-in-psychological-medical-and-pedagogical-applications/virtual-reality-a-new-era-in-surgical-training. Accessed 19 July 2018

Cai S, Wang X, Chiang FK (2014) A case study of augmented reality simulation system application in a chemistry course. Comput Hum Behav 37:31–40

Carleton College (2018) What is gallery walk? 7 May 2018. https://serc.carleton.edu/introgeo/gallerywalk/what.html

Carrard V et al (2020) Virtual patient simulation in breaking bad news training for medical students. Patient Educ Couns 103(7):1435–1438

Castellar EN, All A, De Marez L, Van Looy J (2015) Cognitive abilities, digital games and arithmetic performance enhancement: a study comparing the effects of a math game and paper exercises. Comput Educ 85:123–133

Chien YH et al (2018) Engaging students in using 3D printing technology to enhance cognitive structures and thought processes relevant to engineering design. J Eng Sci Technol 13(Special Issue on ICITE 2018):27–34

Chopra S et al (2016) Met Percy: the CS 221 teaching assistant Chatbot. ACM Trans Graph 1(1)

Christensen G, Steinmetz A, Alcorn B, Bennett A, Woods D, Emanuel EJ (2013) The MOOC phenomenon: who takes massive open online courses and why? SSRN Electron J. https://doi.org/10.2139/ssrn.2350964. Accessed 22 July 2018

Clark D, Tanner-Smith E, Killingsworth S (2016) Digital games, design, and learning: a systematic review and meta-analysis. Rev Educ Res 86(1):79–122

Coronell G et al (2019) Meaningful learning through virtual tutors: a case study. In: 2019 IEEE global engineering education conference (EDUCON), IEEE, pp 276–279

Dalgarno B, Bishop AG, Adlong W, Bedgood DR Jr (2009) Effectiveness of a virtual laboratory as a preparatory resource for distance education chemistry students. Comput Educ 53:853–865

Fain P (2016) Digital, verified and less open. Inside Higher Ed. https://www.insidehighered.com/news/2016/08/09/digital-badging-spreads-more-colleges-use-vendors-create-alternative-credentials. Accessed 10 July 2018

Farhan M, Jabbar S, Aslam M, Hammoudeh M, Ahmad M, Khalid S, Khan M, Han K (2018) IoT-based students interaction framework using attention-scoring assessment in eLearning. Future Gener Comp Syst 79:909–919

Finkelstein ND, Adams WK, Keller CJ, Kohl PB, Perkins KK, Podolefsky NS, Reid S, LeMaster R (2005) When learning about the real world is better done virtually: a study of substituting computer simulations for laboratory equipment. Phys Rev Spec Top Phys Educ Res 1:010103

Ford S, Minshall T (2019) Invited review article: where and how 3D printing is used in teaching and education. Addit Manuf 25:131–150

Gordy CL, Sandefur CI, Lacara T, Harris FR, Ramirez MV (2020) Building the Lac Operon: a guided-inquiry activity using 3D-printed models. J Microbiol Biol Educ 21(1):21.1.28

Hamari J, Shernoff DJ, Rowe E, Coller B, Asbell-Clarke J, Edwards T (2016) Challenging games help students learn: an empirical study on engagement, flow and immersion in game-based learning. Comput Hum Behav 54:170–179

Heller B et al (2016) Conversational agents in second life: Freudbot. Learning and virtual worlds, AU Press, Athabasca University, pp 153–165

Hew KF, Cheung WS (2010) Use of three-dimensional (3-D) immersive virtual worlds in K-12 and higher education settings: a review of the research. Br J Educ Technol 41:33–55

IBM Skills Gateway (2018) What is an IBM Digital Badge? Badges. https://www-03.ibm.com/services/learning/ites.wss/zz-en?pageType=page&c=M425350C34234U21. Accessed 18 July 2018

Johnson L, Adams Becker S et al (2013) NMC Horizon Report: 2013 Higher Education. New Media Consortium

Johnson L, Adams Becker S et al (2014) NMC Horizon Report: 2014 Higher Education. New Media Consortium

Kemper G (2019) Where do wearables fit into the internet of things? IoT for All. https://www.iotforall.com/where-wearables-fit-in-iot/. Accessed 7 Feb 2019

Kim Y, Baylor AL (2015) Research-based design of pedagogical agent roles: a review, progress, and recommendations. Int J Artif Intell Educ 26:160–169

Kinateder M, Gromer D, Gast P, Buld S, Müller M, Jost M, Nehfischer M, Mühlberger A, Pauli P (2015) The effect of dangerous goods transporters on hazard perception and evacuation behavior–a virtual reality experiment on tunnel emergencies. Fire Saf J 78:24–30

Koglbauer I, Kallus KW, Braunstingl R, Boucsein W (2011) Recovery training in simulator improves performance and psychophysiological state of pilots during simulated and real visual flight rules flight. Int J Aviat Psychol 21:307–324

Lakshminarayanan V, McBride AC (2015) The use of high technology in STEM education. Education and training in optics and photonics: ETOP 2015. In: Proc. of SPIE, vol 9793

Lundin M, Rensfeldt AB, Hillman T, Lantz-Andersson A, Peterson L (2018) Higher education dominance and siloed knowledge: a systematic review of flipped classroom research. Int J Educ Technol High Educ 15:20. https://doi.org/10.1186/s41239-018-0101-6

Lynch M (2016) Artificial intelligence and classrooms: will it help or hurt? Education Week, 28 Mar 2016. http://blogs.edweek.org/edweek/education_futures/2016/03/artificial_intelligence_and_classrooms_will_it_help_or_hurt.html. Accessed 10 July 2018

Ma T, Xiao X, Wee W, Han CY, Zhou X (2014) A 3D virtual learning system for STEM education. In: Shumaker R, Lackey S (eds) Virtual, augmented and mixed reality. applications of virtual and augmented reality. VAMR 2014. Lecture notes in computer science, vol 8526. Springer, Cham

Manyika J, Chui M, Woetzel J, Dobbs R (2015) Unlocking the potential of the internet of things. McKinsey Global Institute. https://www.mckinsey.com/~/media/McKinsey/Business%20Functions/McKinsey%20Digital/Our%20Insights/The%20Internet%20of%20Things%20The%20value%20of%20digitizing%20the%20physical%20world/The-Internet-of-things-Mapping-the-value-beyond-the-hype.ashx. Accessed 5 July 2018

McFarland M (2016) What happened when a professor built a Chatbot to be his teaching assistant. Washington Post, 11 May 2016

McNally B, Chipperfield J, Dorsett P, Del Fabbro L, Frommolt V, Goetz S, Lewohl J, Molineux M, Pearson A, Reddan G, Roiko A, Rung A (2017) Flipped classroom experiences: student preferences and flip strategy in a higher education context. High Educ 73:281–298

Moore EB, Chamberlain JM, Parson R, Perkins KK (2014) PhET interactive simulations: transformative tools for teaching chemistry. J Chem Educ 91:1191–1197

National Research Council (2012) A framework for K-12 science education: practices, crosscutting concepts, and core ideas. National Academies Press, pp 7–11

O'Flaherty J, Phillips C, Karanicolas S, Snelling C, Winning T (2015) The use of flipped classrooms in higher education: a scoping review. Internet High Educ 90:85–95

OER Commons (2018) Create OER with Open Author. https://www.oercommons.org/?gclid=C-jwKCAjw7cDaBRBtEiwAsxprXQTQU7C5IMFm8HpaBQfyV99Nnt-e6idcQivzcQXJkW9_VfJuZIqxQhoCaS8QAvD_BwE. Accessed 10 July 2018

Pappas C (2013) 18 free digital storytelling tools for teachers and students. eLearning Industry. Retrieved July 12, 2018 from https://elearningindustry.com/18-free-digital-storytelling-tools-for-teachers-and-students

Pinger CW et al (2019) Applications of 3D-printing for improving chemistry education. J Chem Educ 97(1):112–117

President's Council of Advisors on Science and Technology (2012) Engage to excel: producing one million additional college graduates with degrees in science, technology, engineering, and mathematics. Executive Office of the President. https://obamawhitehouse.archives.gov/sites/default/files/microsites/ostp/pcast-engage-to-excel-final_2-25-12.pdf

Pyatt K, Sims R (2012) Virtual and physical experimentation in inquiry-based science labs: attitudes, performance and access. J Sci Educ Technol 21:133–147

Reece AJ, Butler MB (2017) Virtually the same: a comparison of STEM students' content knowledge, course performance, and motivation to learn in virtual and face-to-face introductory biology laboratories. J Coll Sci Teach 46:83–89

Rincon JA, Costa A, Novais P, Julian V, Carrascosa C (2017) Using emotions in intelligent virtual environments: the EJaCalIVE framework. Wirel Commun Mob Comput 2017. https://doi.org/10.1155/2017/9321463. Accessed 22 July 2018

Shute VJ, Kim YJ (2014) Formative and stealth assessment. In: Spector J, Merrill M, Elen J, Bishop M (eds) Handbook of research on educational communications and technology. Springer, New York

Stewart J, Bleumers L, Van Looy J, Mariën I, All A, Schurmans D, Willaert K, De Grove F, Jacobs A, Misuraca G (2013) The potential of digital games for empowerment and social inclusion of groups at risk of social and economic exclusion: evidence and opportunity for policy. Joint Research Centre, European Commission

Vlachopoulos D, Makri A (2017) The effect of games and simulations on higher education: a systematic literature review. Int J Educ Technol High Educ 14:1–33

von Übel M (2019) The 3D printing materials guide. All3DP. https://all3dp.com/1/3d-printing-materials-guide-3d-printer-material/. Accessed 13 July 2019

Winkelmann K, Keeney-Kennicutt W, Fowler D, Macik M (2017) Development, implementation, and assessment of general chemistry lab experiments performed in the virtual world of Second Life. J Chem Educ 94:849–858

Yin J, Liu H, Wu M, Wu B, Luo M (2014) Carrier flight deck crew training based on immersive virtual simulation and motion capture. Adv Comput Control 59:131–137

Personalized and Adaptive Learning

Deborah L. Taylor, Michelle Yeung, and A. Z. Bashet

Personalized and adaptive learning is currently implemented in a variety of ways. To explore the topic and how it might be applied to STEM disciplines, meet Keisha Johnson, our hypothetical student, and Professor Jones, who is implementing personalized adaptive learning techniques into the classroom. Using their shared story, we will provide definitions, benefits, impact, opportunities, challenges, and future research initiatives while showcasing the choices required by the institution, faculty, and students.

Keisha Johnson, a pre-nursing major, knows that she needs to earn high grades in all of her classes in order to be accepted into a competitive nursing program. Keisha is a scholarship student attending a large university in her state. She was a stellar high school student but is worried because she graduated from an urban school system that has a reputation of having underprepared students for the challenges of college. Her major concern at this point is that she is starting her second year of classes at the University and has enrolled in Human Physiology, a required but very challenging course that has a reputation as a weed-out class. Keisha is also worried that she won't be able to keep up with the work and might not be able to pass a large class with more than 200 students enrolled due to the possible lack of individualized attention from the instructor.

D. L. Taylor (✉)
University of Kansas, Lawrence, KS, USA
e-mail: dtaylor@ku.edu

M. Yeung
Loyola Marymount University, Los Angeles, CA, USA
e-mail: Michelle.Yeung@lmu.edu

A. Z. Bashet
Dallas County Community College District, Dallas, TX, USA
e-mail: azbashet@dcccd.edu

© The Author(s) 2021

J. Ryoo, K. Winkelmann (eds.), *Innovative Learning Environments in STEM Higher Education*, SpringerBriefs in Statistics,
https://doi.org/10.1007/978-3-030-58948-6_2

> *On the first day of class, Professor Anna Jones makes an announcement to the class that they will be using an adaptive, personalized system for their assignments. She explains to her students how the system assesses each individual's prior knowledge before lecture and then directs content delivery and practice questions to create a learning path that is individualized for every student.*
>
> *Keisha is intrigued by Professor Jones' explanation that adaptive learning is a type of personalized learning made possible by the use of computers. The adaptive learning system delivers assessments that evaluate students' prior knowledge and then using artificial intelligence prescribes individualized learning content in a sequential manner that is designed to meet the learning needs of each individual. After hearing how the adaptive and personalized learning system works, Keisha becomes quite excited. This sounds like a game-changer to her. She knows that she is a hard worker and likes the idea of having resources personalized to her learning needs.*

1 Emerging Trends and Pedagogies

1.1 Defining Personalized and Adaptive Learning

In personalized learning, instructional approaches are customized to individual learners. Feldstein (2016) noted that personalized learning is what you do, not how you do it. While it is possible to achieve personalized learning in a classroom, it is quite a daunting challenge to personalize instruction when there are more than a few students in the class. Yet, it is now feasible to bring personalized learning to scale utilizing the affordances of adaptive learning technologies. As a result of the changing face of education, computer technology has greatly affected the pedagogy of higher education. The computer has gone from simply being a conduit for delivering course content to students to one that continuously identifies the learning needs of each student and provides individualized learning paths in real time.

Unfortunately, at this time, in the academic environment, the concept of adaptive learning is nebulous. Cavanagh et al. (2020) emphasize that the lack of clear and consistent terminology throughout the educational arena presents a stumbling block for adaptive learning implementation in higher education. There is an effort to clarify "what" adaptive learning is, and yet the technologies that provide adaptive learning vary so much that it is daunting to even identify what should be included in the definition. Adaptive learning technology provides personalized learning at scale by assessing learners' current skills/knowledge, providing feedback and content, and then constantly monitoring progress by utilizing learning algorithms that provide real-time updates and the necessary tools to improve student learning (Educause Learning Initiative 2017). The Horizon Report (2018) explains that adaptive learning occurs when digital tools and systems are used to create individual learning paths for students based on their strengths, weaknesses, and pace of learning. While this definition is accepted by many, others (Cavanagh et al. 2020; Pugliese 2016) note that the taxonomy is still quite fluid, thus making it difficult to gain consensus on a working definition.

Some adaptive learning systems include profile information from other sources, but most advanced systems create a learning path at the time of interaction with the student. The student's activity profile, learning analytics data, and machine learning then allow the tools to monitor progress and create continuous adjustment to learning paths in real time, in addition to providing personalized scaffolding to promote learning for each student and targeting individualized intervention for improving student success. Adaptive learning environments provide instructors the tools to utilize technologies and data to provide timely feedback on student performance. In describing adaptive learning systems, Mavroudi et al. (2018) quoted Froschl's definition used in his masters' thesis: "in an adaptive system the needs of the learner are assumed by the system itself and, thus, it adjusts its behavior accordingly" (p. 2). In more formal language, Paramythis and Loidl-Reisinger (2004: 182) note that "a learning environment is considered adaptive if it is capable of: monitoring the activities of its users; interpreting these on the basis of domain-specific models; inferring user requirements and preferences out of the interpreted activities, appropriately representing these in associated models; and, finally, acting upon the available knowledge on its users and the subject matter at hand, to dynamically facilitate the learning process."

1.2 Pedagogies of Adaptive Learning

Pedagogy is defined as the methods and practice of teaching (https://www.lexico.com/en/definition/pedagogy). Personalized and adaptive learning are driving a change in higher education from instructor-centered pedagogies to student-centered pedagogies.

Student-centered learning (also known as learner-centered pedagogy) is based on constructivist learning theory and supports student learning by allowing students to make decisions in their learning (Goodman et al. 2018; Hannafin and Land 1997; Wright 2011). Dockterman's (2018) overview of the history of personalized learning informs us that students learn more effectively when instruction is individualized to the learners' needs and that a new pedagogy of personalization recognizes that each student is different. Bringing that pedagogy to scale, however, requires technological intervention that until now has not been available. Identifying students' needs and providing scaffolding for learning are in the heart of adaptive learning platforms. Scaffolding is defined as the support and guidance provided to the learner until the learner can accomplish a task or demonstrate competence independently (Wood et al. 1976).

"Chunking" of content is an online pedagogical approach guided by cognitive information processing (CIP) research that states, in order to reduce cognitive load and enhance learning, information must be broken down into small manageable "chunks" (Mayer 2005). Adaptive learning content delivery is driven by this online pedagogical approach.

Student learning is enhanced when learners are encouraged to evaluate their learning (metacognition), and many adaptive systems employ tools that allow the students to monitor their progress with some systems offering directed metacognitive experiences for the student as he or she moves through the lessons. Adaptive learning technologies promote learning by utilizing the method of retrieval learning which is often called "testing to learn" (Lindsey et al. 2014; Bae et al. 2019; Miller and Geraci 2016; Thomas et al. 2018). An initial assessment of prior knowledge directs the delivery of content and the scaffolding that addresses the learner's needs with continuous assessments in real-time updates and the learning data constantly allowing for the development of an individualized learning pathway for each student. Retrieval learning and metacognition are components of cognitive learning theories.

Online learning design plays a major role in the design of adaptive learning course development. Cavanagh et al. (2020) developed and shared a blueprint for a design framework as well as a pedagogical approach for adaptive learning that can serve as a starting point for institutions considering adopting an adaptive learning technology for course delivery.

1.3   Emerging Technologies

Emerging technologies enabling personalized and adaptive learning include learning analytics, artificial intelligence, machine learning, intelligent tutors, adaptive controls, and robust interactive learning content (Groff 2017; Mavroudi et al. 2018; Murray and Pérez 2015). Adaptive learning platforms provide students a flexible learning environment that can accelerate learning by creating an individualized learning path directed by prior knowledge and continuous assessment of performance. In the 2017 Horizon Report, Adams Becker, Cummins, Davis, Freeman, Hall, and Ananthanarayanan identified adaptive learning technologies as one of the six developments of educational technologies that will have the greatest impact in institutions of higher education. However, the 2019 Horizon Report noted concerns that the expectations for progress were not yet met. Alexander et al. (2019: 35) suggest that the failure to meet expectations may be due to the fact that "Technology tools were felt to be in their infancy, creating a large investment from the institution of time, money, and resources." Yet, the rapid evolution of information processing and Internet technologies enables e-learning to provide personalization, interactivity, media-rich content, just-in-time delivery, and a learner-centered environment where students can take ownership of their learning. Advances in technologies present new opportunities for adapting instruction to individual learning paths. According to the 2017 New Media Consortium (NMC) Horizon Report–Higher Education Edition, "the increasing focus on customizing instruction to meet students' unique needs is driving the development of new technologies."

Adaptive learning technologies can be classified as adaptive learning platforms and adaptive learning programs. The adaptive learning platforms exist as stand-alone systems, where all courseware functionalities are bundled into one robust working unit, whereas adaptive learning programs provide a component that can plug into an existing course in an LMS and deliver the adaptive and personalized experience. Either the adaptive systems can provide content developed by the vendor or the systems can merely serve as a framework for institutions to input their own content.

Pugliese (2016) noted that many promising opportunities exist where adaptive learning technology can enhance student learning and success. Adaptive learning technologies can support content created by a vendor, publisher, or institution. There are pros and cons to each of the options, and it falls to the instructor and institution to sort out which to use based upon institutional resources, availability, quality of the content, and usability of the technology. The adaptive mechanisms vary as well. Currently there are four categories of adaptive learning systems, and all integrate learning objectives, instructional resources, and assessments into modules for learning. The four frameworks (from simplest to most complex) are decision tree adaptive systems, rule-based adaptive systems, advanced algorithm-based adaptive systems, and machine learning-based adaptive systems. A short list of some of the emerging adaptive technologies currently used in higher education includes Smart Sparrow, Knewton, CogBooks, Cengage Mindtap, and Realizeit.

Personalizing learning at scale is made possible by powerful computer processing assessment data, interaction data, and learning behavior data for each student in order to create feedback, scaffolding, and continuous assessment that deliver the individualized learning paths. Adaptive learning systems can inform instructors early in the course, even before the first exam, as the student's activity and performance in the course are continuously monitored and assessed in formative assessments. Information can be sent to the instructor, and early intervention methods can be utilized to catch an at-risk student before it is too late, thus reducing attrition and increasing student success. "Adaptive systems address the fundamentally different levels of prior knowledge, as well as course content progression based on students' skill and outcomes mastery measurement, decreasing faculty load in teaching and remediation to teaching and facilitating" (Pugliese 2016).

2 Use of Personalized and Adaptive Learning in 2026

2.1 Content Presentation

The adaptive learning system develops a learning path for each student and delivers individualized content based upon assessments of performance. One of the more critical requirements is that of delivering the content into "bite-sized" chunks that allow evaluation of bits of knowledge that are then compiled into a learning plan.

The adaptive learning system tracks performance and monitors behaviors during the time of interaction and then delivers the adaptive content in an adaptive sequence individualized for each learner. A robust adaptive learning system offers multiple options for the modality of the content delivery, e.g., text, video, or interactive activities. Some adaptive systems have the capacity to analyze performance and determine if the student learns better by interactive activities, watching videos, or reading and then deliver the content in the ideal format that is associated with that student's improved performance.

2.1.1   Opportunities

In adaptive learning, content delivery is guided by student performance and prior knowledge, thus providing a more personalized experience and promoting student engagement (Dziuban et al. 2016). Kerr (2016) demonstrated that when the adaptive learning systems deliver content in the format determined by assessment to be the best for that student's learning, student learning is enhanced. Additionally, Meccawy et al. (2007) showed that integrating content presentation with learner interaction enhances student engagement and success.

Most adaptive learning systems are accessed through cloud services providing the students with access to course content when they are ready to learn thus enabling student-centered learning, promoting learning autonomy, and encouraging self-regulated learning. Another important component of an adaptive learning system is the ability to provide remediation should a student not have the prior knowledge needed to perform well in the class. This affordance assists underprepared students to gain the knowledge while still in the course, thus not slowing down their forward educational momentum (Dziuban et al. 2016).

2.1.2   Challenges

The adaptive learning technology field is so young that there are no set standards for these tools. This can result in confusion for those wanting to implement this new technology. Some tools offer some adaptive opportunities, but the adaptive options vary considerably from system to system. Many institutions and faculty find it a very real challenge to select the tool that best meets their needs. While many would prefer to develop their own content, the time and costs to do so are a real challenge to institutions and faculty.

2.1.3   Implementation Strategies

The recognized value of adaptive learning courseware in the university system to increase student success has prompted numerous initiatives from the Association of Public and Land-Grant Universities (APLU). One initiative is the formation of the

Personalized Learning Consortium (PLC), a membership organization charged with increasing information on using technology and personalizing learning in order to promote student success. In addition, the APLU, with funding from the Bill & Melinda Gates Foundation, created a grant opportunity for universities to accelerate the adoption of adaptive courseware by public institutions. Eight universities piloted the process of adopting, implementing, and scaling adaptive courseware. One by-product of this initiative was the development of a tool for faculty, instructional designers, and administrators of postsecondary institutions to effectively navigate the market of courseware solutions. Recognizing that the selection of courseware is contextual, with the course context playing a critical role in guiding the selection and implementation of courseware, the Online Learning Consortium, Tyton Partners, and the Gates Foundation developed the Courseware in Context (CWiC) Guide (coursewareincontext.org) to assist institutions in the complex process of evaluating and selecting the appropriate adaptive learning courseware for their courses. SRI International aligned the framework for efficacy research. Education of stakeholders is a major component of the CWiC Guide, and suggestions for implementation are guided by the defined needs of that institution.

2.1.4 Research Questions

Several research questions were proposed at the X-FILEs workshop regarding content presentation and delivery. However, research answering all of the posed questions was found in current literature. This phenomenon supports the premise that adaptive learning is so new that current knowledge is somewhat nebulous. It appears that many, if not most, of the participants did not fully understand the current status of adaptive learning. One of the questions was "How are policies at various institutions determined for selecting and purchasing the content delivery tools?" which is addressed by the APLU's *Implementing Adaptive Courseware Guide* (2017).

Each morning, Professor Jones pulls up the data on student performance on the assignment for that day and identifies topic areas where students performed poorly. She then adjusts her lecture later that day in order to focus on the content topics that the students found most challenging and reduces the amount of time for the content that the students already know to a minimum.

2.2 Interactions and Communications

Interactions and discussions have been shown to be a vital part of student learning. From the time of Plato up to the present, students have depended upon interactions and communication with one another and with their instructors to facilitate learning. The range and depth of interactions and communication and the modality vary greatly across the learning environment. Very few stand-alone adaptive learning systems include opportunities for students to communicate and work collaboratively with other students. Adaptive learning programs that only deliver adaptive learning activities are typically used in conjunction with other teaching modalities in order to facilitate interactions between and among students. The component systems delivering the adaptive learning activities can be used in online, blended, and face-to-face courses to enhance student-centric learning. Interactions and communications in most adaptive learning programs are limited to those between the student and the program, with the exceptions mentioned above offered by only a few robust adaptive learning systems.

2.2.1 Opportunities

Students can choose when they work, and because the assignments are automatically graded, they can receive immediate feedback and scaffolding as needed. A major component of adaptive learning tools is that of feedback and scaffolding. Initially the student is assessed for knowledge, and the results of this assessment guide the delivery of structured feedback and scaffolding as well as new content. Ideally this process offers a student the opportunity to move through the course content at a pace that is determined by his/her prior knowledge as well as the amount of time he/she puts into the learning process. However, typically these adaptive tools are used in conjunction with a regular class (online or face-to-face), but it is not beyond the scope of adaptive learning to develop courses that allow students to complete a course more quickly than a typical semester period. These self-paced courses offer flexibility of scheduling, but do present challenges as explained below. There are some institutions of higher education now offering self-paced courses, but these do not typically consist of adaptive learning software.

The instructor dashboards allow instructors to monitor student progress and identify students who are having problems early in the course so that intervention can occur in a timely manner. The student dashboards allow students to track their progress through the learning materials and promote self-regulation of learning, and some programs even offer metacognitive assessments to enhance the learning experience.

2.2.2 Challenges

For a student to benefit from an adaptive learning system, he or she must initiate interaction with the system, sustain engagement, and successfully complete the assigned activities. A student working in an adaptive learning environment brings a number of behavioral factors into play that affect the interactions with the course content. The internal student-level factors that influence student behavior in an adaptive learning system can be examined from the perspective of motivational theory. It has been shown that engagement, self-determination theory, autonomy, self-regulation, and the level of internal motivation are positively associated with student learning. The action of initiating engagement is driven by motivational factors. Reeve (2012) notes "self-determination theory is unique in that it emphasizes the instructional task of vitalizing students' inner motivational resources as a key step in facilitating high-quality engagement."

As mentioned above, many fully inclusive adaptive software systems do not offer student-to-student interactions nor student-to-instructor interactions. This could result in the student feeling isolated if no compensating adjustments were offered.

2.2.3 Implementation Strategies

Adaptive software programs can be used to offer supplemental learning opportunities for the traditional classroom as well as in the fully online environment. Many of the fully adaptive platforms do not offer student-to-student interaction, and if this type of interaction is to be a part of the course, it is often necessary to utilize the communication opportunities offered by a traditional learning management system.

2.2.4 Research Questions

One research question asked in the workshop was: "Can personalized avatars be used to deliver the feedback and scaffolding to lessen the students' feelings of isolation?" Kim (2012) proposed guidelines for designing virtual change agents (VCAs) (avatars) that would promote student learning needs in a personalized manner in online remedial math courses. The strategies were based on motivational learning theories and the interactions between the student and the VCA.

> *As the semester progresses, Keisha enjoys doing many of the interactive assignments. She likes receiving immediate feedback on her performance and guidance for her study in order to provide efficient use of her time. Keisha even enjoys the metacognitive component of each question and the ability to track her progress through the course. She recognizes the value of the learning resources and takes time to do additional review assignments that the program provides that focus on what she had missed before but limits material that she has already mastered.*

2.3 Learner Activities

Prior knowledge assessments are typically the first activity that the learners encounter in an adaptive learning environment. Results from these assessments guide the development of an individualized learning path that provides personalized content delivery for each student based on their strengths and weaknesses. This content can range from simplistic fill-in-the-blanks to more sophisticated scenario-based learning activities. The importance of engaging students in learning activities is paramount. In many of the adaptive learning systems, the content included in the system replaces a textbook, meaning that all of the learning occurs via the interactions with the adaptive learning system. The learning activities delivered to each student are determined by the results of assessments, thus creating differing learning paths. Students receive different learning opportunities by the adaptive learning system as it analyzes student knowledge and scaffolding needs (Meccawy et al. 2007). Learning activities that engage interactive animations of complex processes (e.g., cell division) with questions interspersed in the activities increase student engagement.

2.3.1 Opportunities

Adaptive learning systems have the capacity to provide targeted information based on individual learning needs (Peter et al. 2010), provide appropriate scaffolding (Raes et al. 2012), and personalize activities based on student responses (Chen 2011; Normandhi et al. 2019; Blair et al. 2016).

Gephardt (2018) studied the effect of utilizing adaptive learning courseware on student performance in an Economics course at Colorado State University as part of an Association of Public and Land-Grant Universities (APLU). She found that the "students who completed low-stakes adaptive assessments outperformed their peers who did not complete the adaptive assignment on easy and moderate questions on the exam that could result in a higher course grade. This suggests that if adaptive learning courseware is integrated as low-stakes assignments then student outcomes improved with relatively little effort on both sides of the instructor and the students" (p. 16).

2.3.2 Challenges

Providing engaging, robust, and applicable learning activities that address the learning outcomes is probably the most challenging aspect of building an adaptive learning experience (excluding building the learning platform). Following that is the ability to develop appropriate feedback and purposeful scaffolding that free the student from needing assistance from the instructor in order to move forward in his or her learning path.

2.3.3 Implementation Strategies

There is often resistance to accept this modality by the faculty due to training needs and time to set up the system (in those that are modifiable). Alignment with course content is challenging since many faculty members determine the content for their own courses. Therefore, it would be important that the system allows for adaptation by the instructor to give him/her ownership. The University of Central Florida modeled an excellent approach in developing and delivering adaptive content in their pilot program (Dziuban et al. 2016). However, it needs to be mentioned that the administration was behind the project and provided a great deal of financing, not only to pay the faculty to develop the content but also to pay graduate teaching assistants and instructional design support to reduce the workload of the faculty. Any institution considering building their own content would do well to explore what the UCF has done.

2.3.4 Research Questions

One question from the 3-day X-FILEs workshop was: "What learning activities are most effective in promoting student engagement and learning?" Linnenbrink and Pintrich (2003) determined that relevant and relatable activities promote student engagement and learning. Additionally, Van Lehn (2011) showed that utilizing an adaptive learning experience promoted student engagement and learning. Therefore, a valuable area of research would be that of examining student behaviors such as persistence, self-regulation, internal motivation, and engagement and see if the adaptive learning systems can capture that data correctly and determine if student learning is affected.

Dr. Jones designed her class so that the adaptive learning system assignments are due the evening before each class begins. This prepares the students for learning to occur during class by remediating those who were lacking in knowledge needed to address new material.

2.4 Assessment

The most effective adaptive learning systems initially assess the student's prior knowledge followed by continuous formative assessment as well as guidance as the student moves through the course content (EdSurge 2016). Dziuban et al. (2018) proposed that adaptive learning acts like a GPS for students. It allows for personalized instruction by altering students' pathways through course objectives. In addition, adaptive learning systems continually assess students' knowledge, guiding them to efficiently and effectively progress through the course.

In conjunction with the creation of a personalized learning path is the requirement of mastery of the material (Gebhardt 2018). Mastery learning requires clear measurable learning objectives, an idea of what mastery of that learning objective entails, learning activities that assess the mastery, and a means of tracking and sharing the information to direct learning. STEM courses depend upon learning new content based upon prior knowledge, and a student moving forward in a lesson without mastering the concepts is a recipe for failure. Adaptive learning provides the assessment tools and evaluation options that can assure mastery of content, not only at the time of new content delivery but periodically checking the continued mastery throughout the learning path.

2.4.1 Opportunities

Adaptive learning courseware can significantly reduce the amount of time a student takes to complete a course when evaluation of prior knowledge shows mastery of that content. The individualized learning pathway provided by the adaptive learning courseware can allow the student to move forward to the next module at a pace that is suitable for him or her. If gaps in knowledge are revealed, a robust adaptive learning platform provides remediation as needed, including appropriate feedback and scaffolding.

Grading open-ended questions such as essays, projects, etc. to assess higher-order thinking skills in didactic curricula of large classes can be onerous. Quite often, instructors use multiple-choice questions (MCQs). However, creating higher-order thinking questions in MCQ format is very challenging and time-consuming because typically a scenario, case study, or fairly complex problem needs to be described and three or four plausible, but incorrect choices created. Aguilar et al. (2006) proposed a computer-based adaptive assessment tool designed to use formative and summative assessments developed by teachers allowing them to use the information to perform real-time evaluations of the learners' levels of understanding. Adaptive learning systems take the process one step further by removing the instructors' tasks of evaluating the information to determine the levels of student understanding as well as providing scaffolding to support the learner where needed, thus freeing them up to have more personal contact with the students.

Formative assessments are efficient for both instructors and students to assess learning in personalized and adaptive learning environments. Instructors can check students' understanding through formative assessment and collect valuable data on student learning and then use that data to modify instruction. Godfrey (2006) found that incorporation of computer-based assessment increased student engagement in learning and instructors believed technology-enhanced assessment tools positively impacted their teaching.

Technology-enabled assessments including assistive technology emerged as potentially powerful mechanisms for measuring growth mindsets or behavioral attributes of students as they engage in the learning process (West 2011). The Information Research Corporation developed an integrated technology platform eTouchSciences to support STEM learning that includes devices that provide multiple forms of feedback, including tactile, visual, and audio, to the student (Thomas 2016). Examples of technology-enabled personalized and adaptive assessment tools used in higher education include Carnegie Mellon University's *Cognitive Tutor*

Software, Pearson's *MyLabs,* McGraw-Hill's *ALEKS Online Tutoring System* and *LearnSmart*, and Australia's *SmartSparrow*.

Intelligent tutoring systems are some of the earliest adaptive learning technologies, and they offer personalized and interactive learning experiences that help students engage in learning more effectively than traditionally based instructional methods. This technology tool enables personalization of learning and evaluation of performance in real time. Intelligent tutoring systems enhance students' engagement and individual learning experiences by providing immediate and adjusted feedback based on each student's learning progression and his or her actions and responses to given questions and lesson activities (Thomas 2016). Adaptive learning systems go one step further to supply scaffolding to add to those interactive experiences in order to enhance learning.

2.4.2 Challenges

Verification of identity is a must. It is very important to ensure that it is really the student doing the work. There are a number of methods to verify identity; however, many are quite costly to the student or institution.

Formative assessments composed of multiple-choice questions, including immediate feedback and scaffolding for learning, can be delivered via learning management systems or adaptive learning tools and can be readily directed by teachers. Formative assessment with feedback and scaffolding becomes more challenging when one incorporates critical thinking and complex problem-solving (Spector et al. 2016: 59). The value of formative assessment cannot be denied, yet the overemphasis on summative assessment has resulted in inadequate resources for formative assessment (Ecclestone 2010; Sadler and Good 2006).

2.4.3 Implementation Strategies

Instructor training and professional development opportunities are paramount to implementing a successful adaptive learning endeavor. The University of Central Florida (UCF) and Georgia State University provide excellent examples of instructor training to promote adaptive learning in their institutions. The UCF developed a support network of instructional designers, technical experts, and content experts to assist faculty in developing courses using adaptive learning, and to further enhance the onboarding experience, they developed a self-paced training course for faculty (Cavanagh et al. 2020). Georgia State offered a comprehensive adaptive learning workshop to foster faculty buy-in and commitment. They found that providing support as needed and building a community of inquiry enhanced the implementation of adaptive learning using the funds received from the APLU Learning Grant (Tesene 2018). Developing and delivering formative assessments require extensive time and resources, yet this is a must for a successful implementation if an institution hopes to build its own adaptive learning content.

2.4.4 Research Questions

A powerful research question is addressed in the literature and is worth mentioning here. That one question was: "What pedagogical tools in adaptive learning environments best promote student engagement (and outcomes)?" Scaffolding is one of the most important pedagogical tools utilized in adaptive learning. Scaffolding is described as the structure of content and feedback in such a manner that the student can work through the lessons without requiring intervention from an instructor, thus promoting autonomy and self-direction. A well-designed adaptive learning platform offers a robust experience with numerous branching options supported by scaffolding to allow students to work independently (Raes et al. 2012).

> *Professor Jones compared the performance of recent students in her course using adaptive learning with that of students from classes before she implemented the active learning component. She was able to quantify a significant increase in student pass rates and a decrease in course attrition after she integrated the learning activities into her course.*

2.5 Co-curricular Activities

Co-curricular activities are learning experiences offered outside the classroom that allow students to expand their interests and perhaps even gain skills that would prepare them for their careers. Co-curricular should not be confused with extracurricular activities like sports and performance opportunities. Co-curricular learning experiences add depth to what is occurring in the classroom, yet are not typically graded. Co-curriculars include leadership skills training, service learning, and study abroad.

2.5.1 Opportunities

Co-curricular activities could easily provide personalization of a student's educational path and the chance to apply and implement lessons learned without interfering with the set academic path the most institutions require. A menu of adaptive learning minicourses could offer students opportunities to gain skills adjacent to their primary courses, thus providing a personalized option of learning opportunities that are not part of the program curriculum.

2.5.2 Challenges

One challenge is to ensure the safety of students participating in co-curricular activities while on and off campus. Another challenge is that many students at large universities have difficulties knowing what opportunities are available to them.

2.5.3 Implementation Strategies

Researchers at the University of Washington discovered that students had difficulty finding co-curricular opportunities because they used many different sources to get information. A recommendation from the study was to address the student context and offer information at several different levels.

2.5.4 Research Questions

Many of the participants at the X-FILEs workshop were unfamiliar with what co-curricular offerings were, and all questions posed were already answered in the literature.

3 Conclusions

Personalized and adaptive learning offers great opportunities to promote student learning, yet it presents a number of challenges that instructors and institutions will need to address in order for an adaptive learning implementation to be successful. The first step is to identify what problems are being addressed in implementing an adaptive learning adoption and after that what "right" adaptive learning approach and system are needed to be determined. Will an add-on component to be used with existing courses within various learning management systems solve the problems that are driving the move to adopt adaptive learning, or is a fully independent adaptive learning platform needed? Questions of compatibility with the current learning management systems will need to be answered if the former option is chosen. Additionally, while decisions to implement an adaptive learning approach are typically made by the institution's administrators, yet without faculty buy-in, it is highly unlikely that a successful outcome will occur. Engaging faculty early in the process is advised coupled with institution-wide engagement increasing the possibility of a successful outcome (Implementing Guide 2017: 2).

At the end of the semester, Keisha is one of the top-scoring students in the class and has received notice that she was accepted into the nursing program. She is hoping that the nursing courses she will need to take will provide learning resources that are adaptive and personalized like the physiology course she just completed. However, adaptive and personalized courses are limited by the resources at a school. And there are still some areas that require additional research to be sure that time and money spent on these courses result in better student outcomes.

References

Adams Becker SA, Cummins M, Davis A, Freeman A, Hall C, Ananthanarayanan V (2017) NMC Horizon Report: 2017 Higher Education Edition. The New Media Consortium, Austin. http://cdn.nmc.org/media/2017-nmc-horizon-report-he-EN.pdf

Aguilar G, Gomez A, Kaijiri K (2006, October) Adaptive teaching and learning using a classroom communication system and an adaptive computer-based assessment tool. In: E-learn: world conference on E-learning in corporate, government, healthcare, and higher education. Association for the Advancement of Computing in Education (AACE), pp 2701–2706

Alexander B, Ashford-Rowe K, Barajas-Murph N, Dobbin G, Knott J, McCormack M et al (2019) EDUCAUSE Horizon Report 2019 Higher Education Edition. EDU19, pp 3–41

Bae CL, Therriault DJ, Redifer JL (2019) Investigating the testing effect: Retrieval as a characteristic of effective study strategies. Learning and Instruction 60:206–214

Blair E, Maharaj C, Primus S (2016) Performance and perception in the flipped classroom. Education and information Technologies 21(6):1465–1482

Cavanagh TL, Chen B, Lahcen RAM, Paradiso JR (2020) Constructing a design framework and pedagogical approach for adaptive learning in higher education: a practitioner's perspective. Int Rev Res Open Distrib Learn 21(1):172–196

Chen LH (2011) Enhancement of student learning performance using personalized diagnosis and remedial learning system. Computers & Education 56(1):289–299

Dockterman D (2018) Insights from 200+ years of personalized learning. NPJ Sci Learn 3(1):1–6

Dziuban CD, Moskal PD, Cassisi J, Fawcett A (2016) Adaptive learning in psychology: wayfinding in the digital age. Online Learn 20(3):74–96

Dziuban C, Moskal P, Parker L, Campbell M, Howlin C, Johnson C (2018) Adaptive learning: a stabilizing influence across disciplines and universities. Online Learn 22(3):7–39

Ecclestone K (2010) Transforming formative assessment in lifelong learning. McGraw-Hill Education, Berkshire

Edsurge (2016) Decoding adaptive. Pearson. Retrieved from https://www.pearson.com/corporate/about-pearson/innovation/smarter-digital-tools/adaptive-learning.html

Educause Learning Initiative (2017) 7 things you should know about Adaptive Learning. https://library.educause.edu/~/media/files/library/2017/1/eli7140.pdf. Accessed 14 Jan 2019

Feldstein M, Hill P (2016) Personalized learning: What it really is and why it really matters. Educause review 51(2):24–35

Gebhardt K (2018) Adaptive learning courseware as a tool to build foundational content mastery: Evidence from principles of microeconomics. Current Issues in Emerging eLearning 5(1):2

Godfrey C (2006) The impact of a classroom communication system on the learning process in eighth-grade special education classes. ARE 5(1)

Goodman BE, Barker MK, Cooke JE (2018) Best practices in active and student-centered learning in physiology classes. AJP Adv Physiol Educ 42(3):417–423

Groff J (2017) Personalized learning: the state of the field & future directions. Center for Curriculum Redesign

Hannafin MJ, Land SM (1997) The foundations and assumptions of technology-enhanced student-centered learning environments. Instr Sci 25(3):167–202

Implementing adaptive courseware: a guide to courseware development, use and evaluation based on the collaborative experience of four public research universities (2017) Personalized Learning Consortium. https://www.aplu.org/library/implementing-adaptive-courseware/File

Kerr P (2016) Adaptive learning. ELT J 70(1):88–93. https://doi.org/10.1093/elt/ccv055

Kim C (2012) The role of affective and motivational factors in designing personalized learning environments. Educ Technol Res Dev 60(4):563–584

Lindsey RV, Shroyer JD, Pashler H, Mozer MC (2014) Improving students' long-term knowledge retention through personalized review. Psychol Sci 25(3):639–647

Linnenbrink EA, Pintrich PR (2003) The role of self-efficacy beliefs in student engagement and learning in the classroom. Read Writ Q 19:119–137

Mavroudi A, Giannakos M, Krogstie J (2018) Supporting adaptive learning pathways through the use of learning analytics: developments, challenges and future opportunities. Interact Learn Environ 26(2):206–220

Mayer RE (2005) Introduction to multimedia learning. The Cambridge handbook of multimedia learning 2:1–24

Meccawy M, Brusilovsky P, Ashman H, Yudelson M, Scherbinina O (2007, October) Integrating interactive learning content into an adaptive e-learning system: lessons learned. In: E-learn: world conference on E-learning in corporate, government, healthcare, and higher education. Association for the Advancement of Computing in Education (AACE), pp 6314–6319

Miller T, Geraci L (2016) The influence of retrieval practice on metacognition: the contribution of analytic and non-analytic processes. Conscious Cogn 42:41–50

Murray MC, Pérez J (2015) Informing and performing: a study comparing adaptive learning to traditional learning. Inf Sci 18:111

Normandhi NBA, Shuib L, Nasir HNM, Bimba A, Idris N, Balakrishnan V (2019) Identification of personal traits in adaptive learning environment: Systematic literature review. Computers & Education 130:168–190

Paramythis A, Loidl-Reisinger S (2004) Adaptive learning environments and e-Learning standards. Electron J e-Learn 2(1):181–194

Peter SE, Bacon E, Dastbaz M (2010) Adaptable, personalised e-learning incorporating learning styles. Campus-Wide Inf Syst 27(2):91–100

Pugliese L (2016) Adaptive learning systems: surviving the storm. EDUCAUSE Review. https://er.educause.edu/articles/2016/10/adaptive-learning-systems-surviving-the-storm

Raes A, Schellens T, De Wever B, Vanderhoven E (2012) Scaffolding information problem solving in web-based collaborative inquiry learning. Comput Educ 59(1):82–94

Reeve J (2012) A self-determination theory perspective on student engagement. In: Christenson S, Reschly A, Wylie C (eds) Handbook of research on student engagement. pp 149–172

Sadler PM, Good E (2006) The Impact of self and peer-grading on student learning. Educational Assessment 11(1):1–31

Spector JM, Ifenthaler D, Sampson D, Yang JL, Mukama E, Warusavitarana A et al (2016) Technology enhanced formative assessment for 21st century learning. J Educ Technol Soc 19(3):58–71

Tesene MM (2018) Adaptive selectivity: a case study in evaluating and selecting adaptive learning courseware at Georgia State University. Curr Issues Emerg eLearn 5(1):6. https://scholarworks.umb.edu/ciee/vol5/iss1/6

Thomas S (2016) Future ready learning: reimagining the role of technology in education. 2016 National Education Technology Plan. Office of Educational Technology, US Department of Education. http://tech.ed.gov/

Thomas RC, Weywadt CR, Anderson JL, Martinez-Papponi B, McDaniel MA (2018) Testing encourages transfer between factual and application questions in an online learning environment. Journal of Applied Research in Memory and Cognition 7(2):252–260

Van Lehn K (2011) The relative effectiveness of human tutoring, intelligent tutoring systems, and other tutoring systems. Educ Psychol 46(4):197–221

West DM (2011) Using technology to personalize learning and assess students in real-time. Brookings Institution, Washington, DC

Wood D, Bruner JS, Ross G (1976) The role of tutoring in problem solving. Journal of child psychology and psychiatry 17(2):89–100

Wright GB (2011) Student-centered learning in higher education. Int J Teach Learn High Educ 23(1):92–97

Multimodal Learning

Bettyjo Bouchey, Jill Castek, and John Thygeson

1 Emerging Trends and Pedagogies

The ubiquitous use of digital technologies is continually reshaping the ways individuals access information, share ideas, and communicate with one another. Doing so requires the nimble use of skills, strategies, and mindsets to navigate, communicate, and collaborate online and across multiple contexts (Leu et al. 2017). These changes have profoundly affected instructional choices in education. In today's education landscape, three key motivations challenge traditional notions of teaching and learning and set forth a strong case for multimodal learning as a critical pedagogy:

1. There exists a proliferation of information in several modes: gestures, visuals, haptics, auditory productions, text-based information, and multimedia. Representing information through different modes, and/or using a combination of modes, can create multiple access points for learning (Bezemer and Kress 2016; Matusiak 2013; Nouri 2018; Sankey et al. 2010).
2. There is an increased opportunity for individuals, including workers and learners, to represent and communicate information using different modes (Blum and Barger 2018; Matusiak 2013).

B. Bouchey (✉)
National Louis University, Chicago, IL, USA
e-mail: bbouchey@nl.edu

J. Castek
University of Arizona, Tucson, AZ, USA
e-mail: jcastek@arizona.edu

J. Thygeson
Drexel University, Philadelphia, PA, USA
e-mail: jrt46@drexel.edu

© The Author(s) 2021
J. Ryoo, K. Winkelmann (eds.), *Innovative Learning Environments in STEM Higher Education*, SpringerBriefs in Statistics,
https://doi.org/10.1007/978-3-030-58948-6_3

3. Students are increasingly diverse (National Center for Education Statistics n.d.).

The academy has long relied upon instructional approaches that favor text-based learning. These approaches are sometimes referred to as factory models of teaching because they include top-down management and favor outcomes designed to meet societal needs and rely on age-based classrooms (Bezemer and Kress 2016; Matusiak 2013; Phuong et al. 2017; Sankey et al. 2010). Researchers posit that presenting information in this one-size-fits-all way does not accommodate the widest possible range of learners' needs (Gee 1996; Phuong et al. 2017). As technologies continue to shape the ways individuals engage in learning, exchange ideas in networks, and communicate in multiple forms, educators must expand their ideas of how, where, under what conditions, and which tools people use to learn in a digital age. The changing nature of student demographics and shifts in the technological landscape call into question traditional ways of teaching and learning.

Technological advances, coupled with considerations of the changing needs of today's learners, call for exploring new directions for multimodal teaching and learning. Multimodal learning (MML) can be defined as "learning environments [that] allow instructional elements to be presented in more than one sensory mode (visual, aural, written)" (Sankey et al. 2010: 853). Multimodality looks at the many different modes that people use to communicate with each other and to express themselves. Multimodal learning is relevant as increases in technological tools and associated access to multimedia composing software have led to the ease of use of many modes in presenting, representing, and responding to information.

MML calls for sensemaking where learners take in information, process, and make personal sense of ideas to form deep learning patterns (Bezemer and Kress 2016; Moreno and Mayer 2007). MML expands the range of choices available in a learning environment so that learners can co-construct learning through their preferred mode while also being challenged by integrating the use of other modes (Nouri 2018; Phuong et al. 2017; Sankey et al. 2010). MML requires a high level of agency (self-discipline) by learners, who must have the metacognition necessary to understand *how* they learn and also *when* to challenge themselves to learn in ways that lie outside their preferred modes. Such sophisticated awareness requires a student to explore their metacognition and requires instructors to explore how assessments are administered and leveraged (Bezemer and Kress 2016; Moreno and Mayer 2007; Phuong et al. 2017; Sankey et al. 2010). MML promotes a greater emphasis on encouraging student voices during the learning process and calls for educators to listen to students' preferences. As such, multiple modes should be used to shape the content, response options, as well as the delivery of instruction (Nouri 2018; Phuong et al. 2017).

As a powerful means to customize and adjust learning strategies to reach diverse learners, MML leverages our technology-rich world, digital tools, and networks and can be used to address Universal Design for Learning (UDL) as well. UDL offers three basic principles that encourage the design of supportive learning environments. These principles call for instructional designs that build in multiple means and modes of (1) *representation* (offering flexible ways to present what we teach and learn), (2) *action and expression* (offering flexible options for how we learn and express what we know), and (3) *engagement* (offering flexible options for generating and sustaining motivation to learn) (Rose and Meyer 2002). Encouraging learners to compose ideas using different modes encourages connections to concepts and illustrates ways to employ the principles of UDL. Moreover, deep learning happens when students are given the opportunity to participate in a range of cognitive and social learning activities that are responsive to their preferences and needs or modes (Bezemer and Kress 2016; Moreno and Mayer 2007). Most importantly, MML serves as well-researched pedagogy that sets a platform for educators needing to shift their emphasis from the delivery of instruction to the engagement of their learner (Bezemer and Kress 2016; Tonsing-Meyer 2013). In this way, MML can be an equalizer for diverse students because the use of different modes encourages diverse ways of communicating. MML can be inclusive of multilingual students, English language learners, and those with learning differences and moreover can set the stage for an engaging, rich, and creative learning experience.

1.1 Aspects of Teaching and Learning Covered in the MML Literature

An MML literature survey was conducted in conjunction with eXploring the Future of Innovative Learning Environments (X-FILEs) project. Several areas of teaching and learning were examined to create a framework to conceptualize new forms of learning in the next five years, in other words in 2026. The areas of teaching and learning spanned (a) content presentation, (b) interactions and discussions, (c) learner activities, (d) assessment and evaluation, and (e) co-curricular learning and activities. The following is an overview of literature pertaining to each of these areas.

1.1.1 Content Presentation

Course content can be presented and delivered in multiple modes, such as visuals, auditory productions, text, and multimedia (Bezemer and Kress 2016; Matusiak 2013; Nouri 2018; Sankey et al. 2010). Through MML, content can be curated, either in advance or dynamically, to accommodate multiple access points. Multiple modes are offered so that students can gravitate toward the modes that best align with their individual styles and needs and to encourage the complementary nature of multimodal content (Nouri 2018; Phuong et al. 2017; Sankey et al. 2010). This section assumes that content is either contained in a learning management system for purely online coursework or blended for hybrid and face-to-face instruction and that multimodal content and delivery can augment traditional lectures.

1.1.2 Interactions and Communications

The use of MML in higher education creates a foundation for flexible, open, and multidirectional interaction and discussions with students, their peers, faculty, staff, and the business and community at large. As students embark on sensemaking and co-construction of knowledge in an MML-based classroom (Bezemer and Kress 2016; Moreno and Mayer 2007), they can use a variety of modes to engage with other stakeholders to shape their understanding. Using different modes, students can spontaneously engage in parallel and complementary dialog to deepen learning, answer questions, and engage in sensemaking of the learning objectives in and around a classroom.

1.1.3 Learner Activities

Jewitt (2008) suggests that how knowledge is represented, as well as the mode and media selected, is a crucial aspect of knowledge construction that makes the form of representation integral to sensemaking and learning, more generally. In other words, the way something is represented shapes *what* is to be learned, the curriculum content, and *how* it is to be learned. In our digital world, our efforts to design instruction must be responsive to the ways we intend learners to use the new knowledge they acquire in the workplace and beyond. MML calls for educators to shift their role from learning designer to learning facilitator (Bezemer and Kress 2016; Tonsing-Meyer 2013). Because MML sets a foundation for a co-construction of knowledge between faculty and students (Bezemer and Kress 2016), learner activities are available just-in-time, based on the student's preferred mode and personalized, drawing in the student's preferences for learning. In addition, the use and integration of different modal representations can reinforce ideas and make learning more memorable.

1.1.4 Assessment

Course assessment strategies must also be multimodal in order to ensure objectivity. Whether formative or summative, all course assessments are designed in a modally agnostic way so that the final product can be produced in the student's preferred mode and preferably in a way that demonstrates the intersectionality and complementary nature of different modes (Jewitt 2008; Nouri 2018; Phuong et al. 2017; Sankey et al. 2010).

1.1.5 Co-curricular

Co-curricular learning refers to activities, programs, and learning experiences that complement, in some way, what students are learning in formal classes, but also serve to expand the relevancy and contextualization of that learning in the real world. Since MML is pedagogical, it does not explicitly reference co-curricular activities or learning. Even so, opening up classroom interactions and discussions into a multimodal environment provides a foundation for co-curricular learning from stakeholders outside of the classroom, such as college and university staff, business and community leaders, friends, and family. Learning from those outside of the classroom should not be overlooked or undervalued.

1.2 Research Questions Answered

The literature on multimodal instruction also answered research questions around design for multimodal instruction and how to use multiple modes to encourage deep learning. Moreover, the literature helped to contextualize many of the early research questions and provided a democratic, pedagogical framework within which to conceptualize content presentation, interactions and discussions, learner activities, assessments and evaluations, and co-curricular instruction in 2026.

1.3 Future Research

As noted in the 3-day X-FILEs workshop, there remain areas that call for future research, specifically in the areas of how MML can impact a student's identity development and self-initiative. Also discussed in the workshop, there remain some practical areas that would benefit from further formalization specific to how institutions of higher education support the technology needed to implement multimodal instruction. More inquiry into the efficacy of some modes over others in terms of deep and sustained learning may also be needed, perhaps as a way of creating synergy across modes as well. Workshop participants also noted the importance of developing a framework for the ethical use of MML that would suggest appropriate use and rules of engagement that support students, faculty, and staff in setting healthy boundaries for immersive learning in an information- and technology-rich society. Further research is also needed to better understand the affordances that multimodal instruction can create for students, especially as it pertains to students with disabilities. Lastly, as co-curricular learning takes hold in the use of multimodal instruction, workshop dialog indicated external validation, perhaps by higher education professionals or other community and business partners of learning credentials, will need to be investigated.

2   Use of Multimodal Learning in 2026

This chapter brings MML into focus through the work of the eXploring the Future of Innovative Learning Environments (X-FILEs) project spanning a series of focus groups, 3-day in-person workshop, and further investigation through a team of researchers, industry partners, and faculty brainstorming on how this democratic pedagogy can shape (a) course content presentation, (b) interactions and discussions within a course, (c) learner activities, (d) assessment and evaluation, and (e) co-curricular aspects of education. Each section outlines opportunities for using MML as a framework, implementation strategies, and key questions that have been answered or left open in the literature.

2.1   Content Presentation

Content presentation has progressed from an educational format in which the instructor is charged with passing on his/her information to students via lecture, recitation, and associated activities into an experience in which information is presented to learners in many ways. While lecture is still used, it is frequently supplemented with additional educational content in additional formats, both in class and away from the classroom.

2.1.1   Opportunities

Several affordances exist in providing multimodal course content, chief among them being the self-directed nature of students selecting content delivered in the most familiar or comfortable mode, which is when the deepest learning occurs (Bezemer and Kress 2016; Moreno and Mayer 2007). Moreover, traditional classroom lectures can be augmented by multimodal content and delivery in the learning management system or can provide the opportunity for classroom time to be used for group- or project-based learning. Augmenting content in a learning management system or the traditional classroom lecture can create an active learning space where students are focused on discovery, rather than consumption of information. Discussed in the X-FILEs workshop, this content created by faculty could be commoditized, whereby students could subscribe to certain modes based on their individual preferences for a small fee, rather than relying on traditional per-credit tuition models. In an online course, MML course content could also spark faculty and course designers to create and curate content in new and innovative ways. Additionally, the multimodal nature of virtual environments such as lab simulations, virtual reality, and augmented reality provide experiential opportunities that would otherwise be very difficult to provide in "real life."

Another opportunity exists to deliver MML course content to those who require assistive technologies. Assistive technology (AT) can be defined as any item, piece of equipment, software program, or product that increases, maintains, or improves the functional capabilities of people with disabilities (Assistive Technology Industry Association 2019). The number of people who could benefit from AT is not insignificant. Currently, 15% of the world's population has some form of disability, and 20% of those people experience significant disabilities. As the world's population continues to grow, so too will the total population of those that experience disabilities; furthermore, this population is aging, and there is an increase in chronic health conditions (World Health Organization 2011).

In addition to reaching individuals who would normally have difficulties accessing education, using these technologies can help educators stay compliant with the law. The Rehabilitation Act of 1973 and the Americans with Disabilities Act addressed workplace support for people with disabilities and were later adapted for use in educational situations (U.S. Equal Employment Opportunity Commission n.d.; U.S. General Services Administration 2018). Providing individuals with disabilities with the opportunity to the same education and in the same time frame as students without disabilities will benefit the education provider by staying compliant with laws and regulations.

2.1.2   Challenges

Workshop participants agreed that chief challenges in designing courses and delivering them to incorporate MML are cost and speed of design. It may be cost- and time-prohibitive to create content in all learning modes. For example, one course module on calculating averages in a statistics class could contain the following content items:

1. A word document with the problem completed by the professor (text-based)
2. A video of the problem being completed as students watch and observe on a virtual whiteboard (gestures, video, and auditory)
3. An interactive, adaptive instructional module where a student answers a series of prompts to solve the problem, with dynamic instruction provided if a wrong answer is inputted or the student indicates they have a question

In addition to the time and cost of curating MML content, faculty may find it difficult to switch between modes while in the "live" classroom environment, suggesting more of a need to augment the lecture through MML content in a learning management system as well. Faculty and instructional designers may also find it difficult to find or create content in several modes, especially if they fall outside of their individually preferred modes.

Another challenge of using MML in course content is relying on a student's self-awareness and self-direction to explore and represent content in their preferred mode or combination of modes. Accessing all course content in one mode, even if preferred, might not be the most effective approach as combining modes may serve as reinforcement and encourage connections. There may also be implications with respect to accessibility where some modes might be richer than others. For example, a student who only accesses materials in a preferred mode may not benefit from deeper learning utilizing content presented in different modes or ones that incorporate many modes such as virtual reality.

2.1.3 Implementation Strategies

Incorporating MML in course content and delivery of this type of instruction in "real time" by a faculty member requires a symbiotic relationship with the student so that the faculty member can match instruction to the learning needs of their students. In thinking through online courses, curating content requires examining instructional design frameworks that guide the thoughtful assembly of course content. The seminal and informative ADDIE model of design, which includes the elements of *a*nalysis, *d*esign, *d*evelopment, *i*mplementation, and *e*valuation (Molenda 2003), can still be used, perhaps with iterations derived from rapid prototyping.

2.1.4 Research Questions

Literature on MML clearly answers the research questions set forth by the X-FILEs team in this area of how to choose the best mode for a subject or student group. Through its very nature, MML provides content that addresses all modes regardless of subject matter or student group characteristics.

As Juan accesses his online class for his Introduction to the Fundamentals of Science, he is faced with many choices that he will navigate throughout the term. He has a selection of content available to him to support his first week of learning, including chapters from an online textbook, a series of videos and podcasts, and a set of keywords he can use to locate his own or additional content—he is also asked to register for his asynchronous virtual reality classroom where he can access his book, the videos, and podcasts and perform and post his research. His university has obtained a grant to fund the augmented course design, and Juan has a low-cost technology fee that he pays each term to offset the cost of the virtual classroom licensing, hosting, and support to the institution. The virtual classroom is where the majority of his course will actually take place throughout the week, though he will have several opportunities to participate in the learning "live" or through other modes as well. To provide the students with a sense of familiarity, the virtual classroom is designed just as a school is; each room corresponds to a different subject or module. Juan navigates his avatar into each room to view content and participate in class as though he was sitting in a physical classroom.

2.2 Interactions and Communications

Interactions and discussions around academic coursework have evolved from episodic, in-class lectures and peer dialog to full-scale, immersive, and 24/7 opportunities for dialog and deep learning. While in-class lecture and dialog remain central ways to build foundational learning, the co-curricular opportunities for interactions and discussions serve as a platform for more immersive learning.

Using the principles of UDL (Gordon et al. 2014) and frameworks that encourage the wide use of MML, students, peers, faculty, staff, and community/business stakeholders can engage in open and flexible and dialogic interactions and discussions. Dialogic interactions create space for shared leadership in learning in which learners and instructors become co-inquirers; in this way, a more egalitarian classroom environment can emerge, devoid of traditional teacher/student power dynamics. Together, they collaboratively engage in creating and evaluating new interpretations in order to gain insight into the world, themselves, and each other (Reznitskaya 2012). Learners can and should be viewed as active drivers of their own learning (Bezemer and Kress 2016; Sankey et al. 2010). In this way, students can choose multiple modes within which to engage in learning and receive just-in-time or real-time information to help contextualize their learning. Providing an opportunity for multidirectional modes of communication enables students to be resourceful and frees up faculty time as well.

2.2.1 Opportunities

As students shift from consuming knowledge to actively driving and constructing it, they can enhance their understanding by using different modes to represent their understanding. Interacting across modes frees students to think more creatively and integrate knowledge faster. By opening up meaning-making (Postman and Weingartner 1969), the process by which people interpret situations, events, objects, or discourses, to expand communication, MML may also empower students to generate meaning and make sense of themselves, their experiences, and their relationships through their learning. Moreover, MML affords students who may have trouble communicating in a particular mode with the opportunity to express themselves in another. In this more flexible and dynamic environment, students with disabilities can freely operate within their most capable mode. MML also provides new avenues for student collaboration, beyond the confines of the traditional classroom or learning management system.

2.2.2 Challenges

Despite the benefits and possibilities that MML can make possible, there are also challenges introduced. For example, the use of images to convey meaning requires faculty to develop visual literacies required to infer how others may interpret their creations. Not all viewers walk away with the same meaning from visuals, and thus other modes may need to be integrated to convey a common meaning. Identified in the workshop, another challenge may be introduced when initializing interactions exclusively via technology among learners who come from different cultural backgrounds. In the absence of context clues that come from facial expressions, body language, and gestures, miscommunications may occur. Pairing MML strategies with opportunities to connect through two-way videoconference technologies and to openly dialog about different communication norms across cultural groups may support overcoming these challenges. A final challenge with MML identified in the workshop centers around limiting opportunities for face-to-face communication due to technological dependency. Relying solely on technologies for bridging distances between learners may in fact create more distance and feelings of isolation. MML is intended to provide options for learning and instruction and does not intentionally aim to create distance-only options. Discussing ideas face-to-face, in combination with the use of different modes of meaning-making, can provide flexible options for learning engagement. We aim to support educators in their implementation of MML in support of designing learning interactions of all kinds, including face-to-face.

Another challenge in the use of MML in discussions and interactions is the level of ability in technology use among faculty, staff, and students. The challenge stems from the concept that today's students are digital natives, while most of today's teachers are digital immigrants (Bezemer and Kress 2016; Blum and Barger 2018; Kirschner and De Bruyckere 2017; Lambert and Cuper 2008; Moreno and Mayer 2007; Phuong et al. 2017; Picciano 2009; Prensky 2001; Sankey et al. 2010; Wang et al. 2014). A less rigid definition of the digital natives/immigrant is now used in which the technology exposure is more a factor of a variety of reasons, such as demographics, opportunity, social influence, cognitive knowledge, and socioeconomic status (Kesharwani 2020). A key concept to take from these studies on digital natives/immigrants is that a primary focus should be placed on reaching across the technological divide between varying technology levels when designing course content. Relatedly, it is important to think through access to technology among different members of the population and how this relates to the value of MML to discussions and interactions.

Lastly, as technology evolves, communication channels must be evaluated for privacy and security. Workshop attendees stressed that it is critically important for faculty and students to know what modes and channels may be open for public viewing. In fact, faculty professional development on tool privacy and accessibility will be important and ever-changing year to year or perhaps term to term as the speed of technological advances and tool creation continues to increase. At the same time, students need orientation to tool privacy and accessibility and experience with the types of tools the university provides, as opposed to ones they are not. There may be tools that do not meet privacy or accessibility standards that the university would not support. The choice of tools and the benefits and drawbacks, should be communicated clearly to students.

2.2.3   Implementation Strategies

To motivate students to interact beyond the traditional faculty-student relationship, faculty can provide for multimodal communication either in advance, while a course is in progress, or following learning activities to promote reflection (Dewsbury 2019). For example, a classroom meeting or online discussion question could have a companion virtual world classroom where students interact in real time through avatars or a backchannel for a face-to-face meeting where students can contextualize the lecture with peers or other educational stakeholders. Course participation could also be linked to time-limited assignments in the workplace or community in person, through videoconference, or virtual networking to encourage idea-sharing and real-world application of learned content.

2.2.4   Research Questions

While research in MML is not entirely new, research that addresses challenges identified in the workshop concerning second language learners are an important direction to pursue. Offering multimodal texts is often beneficial for second language (L2) learners as multiple modes can scaffold understanding (Plass et al. 1998; Royce 2002). This is because, as Farias and Abraham (2017) state:

> multimodal texts more effectively support second language reading by providing input that caters to different learning styles and that they are familiar, authentic, and contextualized to the learners' lives. Moreover, these texts facilitate learners' meaningful interaction not only intratextually, by exploring the text/image semiosis, but also intertextually, by allowing readers to become literate in the different genres that are constructed multimodally.

Thus, while L2 learning is supported by the use of multimodal texts, it can also develop learners' broader understanding of concepts. Examining identity and literacy development surrounding the creation of multimodal texts both within and beyond course objectives is an fruitful area in need of additional research. These directions can further support MML for L2 learners across fields. While MML seems to indicate there are benefits associated with the use of multiple modes of communication for students, less is understood about the benefits of students' choices; research into this area is still needed.

In the virtual classroom, Juan can engage through phone, text, email, instant messaging, or other means with his professor, a group of scientists in industry and other colleges, as well as all his peers and advisors. Professor Aranda also actively engages in the virtual classroom and spends time cultivating personalized relationships with each student—in addition to maintaining the relationship with the community of scientists and supporting the classroom. During the Tuesday evening session, she carefully plans a short didactic lecture that is then supplemented with a series of small group virtual breakout rooms facilitated by other scientists where students and scientists work through the theory and discuss ways to demonstrate mastery of the learning objectives for the week.

2.3   Learner Activities

Learner activities can take many shapes and forms and may connect resources available through coursework to online and community resources. The choices faculty make include multimodal options that build toward applying what is learned within and beyond the classroom. However, choices often depend on individual faculty and the pedagogical disposition of a department, college, or university. In this way, learner activities can be varied and may involve multiple formats and response options that call for different ways to illustrate learning.

MML calls for educators to shift their role from lecturer to learning facilitator and designer (Bezemer and Kress 2016; Tonsing-Meyer 2013). MML encourages fluidity between accessing information using multiple modes and representing knowledge through the integration of multiple modes. This approach encourages personalization as well as creativity and calls for a co-construction of knowledge between faculty and students (Bezemer and Kress 2016). In an MML environment, learner activities can be more collaborative, designed to require real-world problem-solving, and created to apply knowledge based on a compelling purpose for learning.

2.3.1   Opportunities

In shifting the focus from lecturer to learning designer or facilitator, faculty are free to work with individual students to understand their current state of learning and to work collaboratively on learner activities that will enhance their mastery. Students are thus empowered to find the best ways to integrate the subject matter alongside faculty. Faculty can also learn from student-designed learner activities when they use modes and tools unfamiliar to the faculty.

2.3.2   Challenges

When students drive their own learning, they are able to benefit from the multiple modes that are available to them. Working between and across modes may serve to support deeper learning and improve retention of key concepts. MML requires a high degree of agency, and provides opportunities for learners to select a focus or direction for themselves. This requires that learners develop the metacognition necessary to understand *how* they learn and also *when* to challenge themselves (Bezemer and Kress 2016; Moreno and Mayer 2007; Phuong et al. 2017; Sankey et al. 2010). The time needed for faculty to engage with students on learner activity design may be prohibitive in larger courses. It also may be difficult to objectively assess learning when activities are varied across a student group.

2.3.3   Implementation Strategies

Learner activities in an MML environment are student-driven, based on preferred modes. Still, faculty could develop a range of learner activities, mapped to course learning objectives, that can be completed in several modes to enable students to select an appropriate activity. These activities may be experiential and hands-on or operate in a virtual environment. For example, an assignment either could be open-ended in terms of how the learning is demonstrated such as a document, a presentation, or a video or could have a range of possible formats the students could choose from.

2.3.4   Research Questions

Research questions in this area rest on the opportunity that exists to operationalize MML-based learner activities to promote student engagement. The MML literature clearly supports the shift from subject matter delivery to that of engagement, as well as the positive benefits of this. Workshop participants aptly pointed out that it will also be important to compare MML-based learner activities to other types of pedagogical frameworks for efficacy.

All of Juan's learning activities are "hands-on" in the virtual classroom and include a series of science experiments that he can perform in the virtual space, at home (where he can record and post in the virtual classroom), or at a series of physical labs his university has partnered with. Throughout the course, Professor Aranda is working to ensure that the course content, learning activities, and support from other scientists and peers all work together to further his individual learning. Professor Aranda is aware that Juan's preferred modes of learning are in visual and auditory realms so she takes time to ensure that he is challenged to incorporate writing and presentation skills, as well as to spend time in the virtual space to drive the intersectionality of multimodal learning and to create more strength in other modes of learning and information consumption.

2.4   Assessment

Traditional means of assessment and evaluation have often revolved around tests and written assignments. Over time, educators have embraced new ways of assessing competency through action projects, multimedia presentations, and other types of practical application strategies. MML can be a helpful framework in the assessment strategy in a course or classroom environment that frees up the faculty and students for more creative ways to show mastery. No matter the type of assessment, formative, summative, or performance-based, designing assessments that invite different response options may free students from a specific way of illustrating their knowledge. In this way, through MML options, learners are encouraged to produce demonstrations of their learning in modes that they have an affinity for. Moreover, providing suggestions for multiple ways to illustrate learning may open the doors to creativity and deeper levels of learning (William and Flora Hewlett Foundation 2016).

2.4.1 Opportunities

By designing assessments so that they can be completed in multiple modes, students can draw from their modally-driven learning to produce artifacts in the mode that they are most comfortable and familiar with. By reducing the cognitive load associated with using an unfamiliar or uncomfortable mode, learning and mastery can be more easily conveyed. Students could avail themselves of technological advancements to capture their learning as it evolves, ostensibly faster than higher education curricular design cycles. In addition, by opening up the classroom co-curricularly as discussed in section 2.5, assessments can be conducted by stakeholders outside of the classroom walls, providing helpful feedback from employers, community partners, and others that can help shape the student's learning in ways that may not be within the scope of what a faculty member knows or has experienced, personally. Ultimately, the opportunity to have multiple stakeholders providing evaluation and feedback affords faculty with more data to gauge learning and to plan to reduce gaps and increase levels of mastery. With more and consistent data streams through multimodal evaluation, the possibility for machine learning to aid in faster and more efficient grading also presents itself.

2.4.2 Challenges

Students might not choose a mode that effectively demonstrates their learning, or they may select a mode that the faculty member is not familiar with. Diverse outputs may challenge the use of rubrics or programmatic assessment plans that rely upon standardized artifacts or other uniform, objective measures given to students. Moreover, it may be very time-consuming for a faculty member to grade assessments produced in many modes because mastery may "look" different based on the mode in which it was produced. Students may also choose to demonstrate their learning in a mode that lacks adequate privacy or accommodations for students with disabilities—presenting concerns around assisting their selection of modal delivery as part of the assessment directions.

2.4.3 Implementation Strategies

In higher education, a culminating project or final assignment often requires learners to synthesize, analyze, and create content using a variety of modalities. Such projects, when closely aligned with learning outcomes, support relevant and authentic assessment that leverages digital communication strategies. An emerging practical model designed specifically for multimedia instructional strategies, CASPA, contains five components: *c*onsume, *a*nalyze, *s*caffold, *p*roduce, and *a*ssess. This model can be used to promote curriculum-based integration of multimodal projects for assessment in higher education courses. The CASPA framework for assignment design asks students to iteratively attempt and create culminating assessments throughout a course (Blum and Barger 2018). The framework allows students to experiment with different modes as they complete a culminating assignment. Faculty feedback and direction throughout the term could be used to enable learners to choose the best modal demonstration of their learning by the time the end product is due. Faculty will design assessment artifacts that are open-ended and driven by course learning objectives. No formal description of the end products or artifacts is provided to the student, though it may be appropriate to provide options to help students choose effective outputs or ones that protect the student's privacy and that are accessible to all students, regardless of their abilities. Faculty may also consider creating standardized metrics that would allow for cross-modal comparisons of learning products illustrating that there are multiple ways to achieve course aims.

2.4.4 Research Questions

Research questions in this area center on the affordances of some modes over others, and this remains an area for further research as identified by workshop participants; perhaps some modes are more effective than others in the demonstration of learning. Furthermore, faculty modal preference is rarely discussed in the literature. Faculty who are evaluating artifacts outside their modal preference may not be able to objectively assess them due to the cognitive load required to adapt to new or unfamiliar modes.

Juan and his peers are encouraged to locate and post content that aligns to each week's learning objective through a framework of information validation provided by the university. Students are encouraged to comment on these new content postings as a way of sensemaking and co-construction of knowledge. Experiments are the main source of assessment and evaluation in his course, but Juan also has several opportunities to identify and capture scientific theories in his everyday life through a system of digital storytelling. This can include taking a picture and posting about an experiment, recording audio or video about seeing science in nature, or providing a summary of a scientific experience through video or a text-based summary. Professor Aranda awards points to every learning capture that he is willing to submit in the course; in fact, all of the assessments in the course are documented as merely a list of learning objectives so that each student can demonstrate mastery through modes that they have an affinity for or modes that lend themselves to the objective being assessed.

2.5   Co-curricular Activities

While the idea that learning does not always take place in the traditional classroom setting has been accepted for many years (Dewey 1997), educators and students are now embracing the concept. Co-curricular learning refers to activities, programs, and learning experiences that complement, in some way, what students are learning in formal classes, but also serve to expand the relevancy and contextualization of that learning in the real world. Co-curricular contexts enable the learner to interact with material outside the classroom in such a way that they acquire new knowledge or apply a skill from one area to a new area. Increasingly, institutions of higher education are embracing co-curricular learning and its benefits to their students (Stirling and Kerr 2015; Turrentine et al. 2012).

2.5.1   Opportunities

By broadening learning to extend beyond the academy, students can build professional networks, enhance their learning through multiple perspectives, and more quickly build workforce-ready skills. Through co-curricular access to other educational stakeholders, students can engage and consume new learning in the mode they most prefer. Students could also participate and drive change in their community by working directly with business and community leaders to couple their learning with practical applications.

These co-curricular activities could take many forms, and in one scenario, educational stakeholders such as a venerable scientist or engineer would be purposefully contracted by the institution to participate in the virtual world associated with the course to:

1. Host weekly informational sessions via web conferences, virtual world environments, or even via chat
2. Make himself/herself available for text- or video-based question/answer sessions on the week's subject matter from the classroom
3. Use virtualized laboratories for student observation
4. Provide feedback and insight into informal or formal assessment of the subject matter

Each scenario further extends the student's learning, infused with the relevancy and currency of a working professional in their field of study.

2.5.2   Challenges

Many institutions of higher education do not have formal connections with the private sector that underrepresented students' need to benefit from co-curricular learning. Faculty may also be challenged with assessing which learning should happen outside the classroom and what should be inside the classroom (face to face or online), may struggle with how to integrate or sequence this type of learning with classroom-based learning. Federal credit hour definitions may also limit co-curricular learning or make it difficult to "count" it, so it would be additive, perhaps contributing to cognitive overload for students. Validation of co-curricular learning may be difficult to assess as well.

2.5.3 Implementation Strategies

Faculty can design course learning objectives tied to business and community interaction, where the choice of modalities would be driven by a student's individual preferences. For example, a course learning objective could read "Students will consult with a scientist to discuss what types of careers a science graduate can pursue upon graduation." Faculty could design course content and set the stage for interactions and discussions that engage educational stakeholders from outside the walls of the classroom.

2.5.4 Research Questions

Research questions in this area center on the benefits and consequences of students' learning outside of the classroom. MML literature is seemingly quiet in this area, though it was a theme discussed in the workshop. There does remain a risk to the fidelity, meaning the degree to which application of MML can be replicated, across iterations of co-curricular learning programming. As co-curricular learning increases, capturing the outcomes and related learning that stems from MML will need to be developed and well understood by faculty and students as well.

Each Tuesday night, Juan's whole class, his professor, and a group of scientists from all over the world gather to sit in the virtual classroom to discuss vital scientific theory that will enable him to be more successful when he is attempting his experiments. These same scientists are working with Juan and Professor Aranda throughout the term to customize the learning and to be sure Juan is engaged in the most up-to-date practice. Juan is developing a strong network of future colleagues as he interacts with these external educational stakeholders.

3   Conclusion

While there are clear opportunities and affordances offered through the use of MML in the classroom, there are also inherent limitations to MML, particularly related to cognitive load. Too many modes and too much information can overwhelm learners

(Moreno and Mayer 2007; Sweller et al. 1998). Furthermore, fundamentally shifting traditional teaching would require intentional, responsive, and on-going professional development so that teachers can develop multimodal literacies. One must also consider the cost and complexity (Bezemer and Kress 2016) of curating learning environments so that learners can choose from all of the varied types of modes. Though challenges do exist in the use of MML, the opportunities and affordances that exist through incorporating facets of MML in classroom teaching and learning far outweigh the challenges. Moreover, as student demographics continue to evolve into more and more diversity and technology continues to reshape learning interactions, instruction will inherently take on a multimodal nature.

In the end, Juan finishes his Introduction to the Fundamentals of Science course with a deep understanding of scientific theory and its application in the real world. He also leaves the course with a newly formed group of mentors which includes Professor Aranda and the scientists who supported the course. He has also formed deep bonds with his peers and plans to connect with them in future courses. Lastly, he leaves the course with a new sense of how improving his writing and presentation skills can provide value not only in deepening his learning but in how he presents himself to the world outside of the university.

References

Assistive Technology Industry Association (2019) What is AT? https://www.atia.org/at-resources/what-is-at/

Bezemer J, Kress G (2016) Multimodality, learning and communication: a social semiotic frame. Routledge, New York

Blum M, Barger A (2018) The CASPA model: an emerging approach to integrating multimodal assignments. J Educ Multimedia Hypermedia 27(3):309–321

Dewey J (1997) Experience and education. Simon & Schuster, New York

Dewsbury B (2019) Teaching with technology in higher ed: start with relationship-building. EdSurge. https://www.edsurge.com/news/2019-01-02-teaching-with-technology-in-higher-ed-start-with-relationship-building

Farias M, Abraham P (2017) Reading with eyes wide open: reflections on the impact of multimodal texts in second language reading. Íkala 22(1):57–70. http://dx.doi.org.ezproxy4.library.arizona.edu/10.17533/udea.ikala.v22n01a04

Gee JP (1996) Discourses and literacies. In: Luke A (ed) Social linguistics and literacies: ideology in discourses, 2nd edn. Taylor & Francis, London, pp 122–148

Gordon D, Meyer A, Rose D (2014) Universal design for learning: theory and practice. CAST Professional Publishing

Jewitt K (2008) Multimodality and literacy in school classrooms. Rev Res Educ 32:241–267

Kesharwani A (2020) Do (how) digital natives adopt a new technology differently than digital immigrants? A longitudinal study. Inf Manag 57(2):103170. https://doi.org/10.1016/j.im.2019.103170

Kirschner PA, De Bruyckere P (2017) The myths of the digital native and the multitasker. Teach Teach Educ 67:135–142. https://doi.org/10.1016/j.tate.2017.06.001

Lambert J, Cuper P (2008) Multimedia technologies and familiar spaces: 21st century teaching for 21st-century learners. Contemporary issues in technology and teacher education 8(3). https://www.citejournal.org/volume-8/issue-3-08/current-practice/multimedia-technologies-and-familiar-spaces-21st-century-teaching-for-21st-century-learners/

Leu DJ, Coiro J, Kinzer C, Castek J, Henry LA (2017) New literacies: a dual level theory of the changing nature of literacy, instruction, and assessment. Special issue teaching and learning in the 21st century. J Educ 197(2):1–18

Matusiak KK (2013) Image and multimedia resources in an academic environment: a qualitative study of students' experiences and literacy practices. J Am Soc Inf Sci Technol 64(8):1577–1589. https://doi.org/10.1002/asi.22870

Molenda M (2003) In search of the elusive ADDIE model. Perform Improv 42(5):34–36

Moreno R, Mayer R (2007) Interactive multimodal learning environments. Educ Psychol Rev 19(3):309–236

National Center for Education Statistics (n.d.) Table 203.60 in Digest of Education Statistics. https://nces.ed.gov/programs/digest/d17/tables/dt17_203.60.asp?

Nouri J (2018) Students multimodal literacy and design of learning during self-studies in higher education. Technol Knowl Learn 24:683. https://doi.org/10.1007/s10758-018-9360-5

Phuong AE, Nguyen J, Marie D (2017) Evaluating an adaptive equity-oriented pedagogy: a study of its impacts in higher education. J Effect Teach 17(2):5–44

Picciano AG (2009) Blending with purpose: the multimodal model. JALN 13(1):7–18

Plass JL, Chun DM, Mayer RE, Leutner D (1998) Supporting visual and verbal learning preferences in a second language multimedia learning environment. J Educ Psychol 90:25–36

Postman N, Weingartner C (1969) Teaching as a subversive activity. Delacorte Press, New York

Prensky M (2001) Digital natives, digital immigrants. On the Horizon 9(5)

Reznitskaya A (2012) Dialogic teaching: rethinking language use during literature discussions. Read Teach 65(7):446–456

Rose D, Meyer A (2002) Teaching every student in the digital age. ASCD, Alexandria. http://www.cast.org/teachingeverystudent/ideas/tes/

Royce T (2002) Multimodality in the TESOL classroom: exploring visual-verbal synergy. TESOL Q 36(2):191–205

Sankey M, Birch D, Gardiner M (2010) Engaging students through multimodal learning environments: the journey continues. In Proceedings of ASCILITE—Australian Society for Computers in Learning in Tertiary Education Annual Conference 2010, pp 852–863

Stirling AE, Kerr GA (2015) Creating meaningful co-curricular experiences in higher education. J Educ Soc Policy 2(6). http://jespnet.com/journals/Vol_2_No_6_December_2015/1.pdf

Sweller J, van Merrienboer JG, Paas FC (1998) Cognitive architecture and instructional design. Educ Psychol Rev 10(3):251–296

Tonsing-Meyer J (2013) An examination of online instructional practices based on the learning styles of graduate education students. Q Rev Dist Educ 14(3):141–149

Turrentine C, Esposito T, Young MD, Ostroth DD (2012) Measuring educational gains from participation in intensive co-curricular experiences at Bridgewater State University. J Assess Inst Effect 2(1):30–54

U.S. Equal Employment Opportunity Commission (n.d.) The Rehabilitation Act of 1973. https://www.eeoc.gov/laws/statutes/rehab.cfm

U.S. General Services Administration (2018, November) IT Accessibility Laws and Policies. https://www.section508.gov/manage/laws-and-policies

Wang S, Hsu H, Campbell T, Coster D, Longhurst M (2014) An investigation of middle school science teachers and student use of technology inside and outside of classrooms: considering whether digital natives are more technology savvy than their teachers. Educ Technol Res Dev 62(6):637–662. https://doi-org.ezproxy2.library.drexel.edu/10.1007/s11423-014-9355-4

William and Flora Hewlett Foundation (2016) What is deeper learning? http://www.hewlett.org/programs/education/deeper-learning/what-deeper-learning
World Health Organization (2011) World report on disability. WHO Press, Switzerland. http://whqlibdoc.who.int/publications/2011/9789240685215_eng.pdf?ua=1

Cross Reality (XR): Challenges and Opportunities Across the Spectrum

Cindy Ziker, Barbara Truman, and Heather Dodds

1 Emerging Trends and Pedagogies

Cross Reality (XR) refers to a group of emerging technologies such as virtual reality (VR), augmented reality (AR), and virtual worlds (VWs) that involve the use of 3D models/simulations across physical, virtual, and immersive platforms. The path to optimizing the use of XR in education is not always easy to navigate. However, with adequate support, XR has the potential to help faculty and students transcend the boundaries of the classroom by providing new types of environments for presenting and delivering instructional content and creating learning experiences with the power to develop unique communities of inquiry and practice.

Throughout this chapter, we will explore scenarios populated by a typical future student named Andi and share examples of her engagement with Cross Reality throughout her educational career. These scenarios are designed to illustrate what lies ahead in education, as these technologies become more and more ubiquitous, allowing students like Andi to move seamlessly within the Reality-Virtuality Continuum.

C. Ziker (✉)
Ziker Research, San Jose, CA, USA
e-mail: cindyziker@yahoo.com

B. Truman
University of Central Florida, Orlando, FL, USA
e-mail: btruman@ist.ucf.edu

H. Dodds
Independent Researcher, New York, NY, USA
e-mail: heatherelizabethdodds@gmail.com

1.1 Definitions

For the purpose of this chapter, Cross Reality or XR refers to technologies and applications that involve combinations of mixed reality (MR), augmented reality (AR), virtual reality (VR), and virtual worlds (VWs). These are technologies that connect computer technology (such as informational overlays) to the physical world for the purposes of augmenting or extending experiences beyond the real. Especially relevant to the definition of XR is the fact that this term encompasses a wide range of options for delivering learning experiences, from minimal technology and episodic experiences to deep immersion and persistent platforms. The preponderance of different terms for slightly different technologies indicate that this is a growth area within the field. Here we provide a few definitions of these technologies.

MR—Mixed reality refers to a blend of technologies used to influence the human perception of an experience. Motion sensors, body tracking, and eye tracking interplay with overlaid technology to give a rich and full version of reality displayed to the user. For example, technology could add sound or additional graphics to an experience in real time. Examples include the Magic Leap One and Microsoft HoloLens 2.0. MR and XR are often used interchangeably.

AR—Augmented reality refers to technology systems that overlay information onto the real world, but the technology might not allow for real-time feedback. As such, AR experiences can move or animate, but they might not interact with changes in depth of view or external light conditions. Currently, AR is considered the first generation of the newer and more interactive MR experiences.

VR—Virtual reality, as a technological product, traces its history to approximately 1960 and tends to encompass user experiences that are visually and auditorily different from the real world. Indeed, the real world is often blocked from interacting with the virtual one. Headsets, headphones, haptics, and haptic clothing might purposely cut off all input except that which is virtual. In general, VR is a widely recognizable term, often found in gaming and workplace training, where learners need to be transported to a different time and place. VR experiences in STEM often consist of virtual labs or short virtual field trips.

VW—Virtual worlds are frequently considered a subset of VR with the difference that VWs are inherently social and collaborative; VWs frequently contain multiple simultaneous users, while VRs are often solo experiences. Another discrimination between virtual reality and virtual worlds is the persistence of the virtual space. VR tends to be episodic, with the learner in the virtual experience for a few minutes and the reality created within the experience ends when the learner experience ends. VWs are persistent in that the worlds continue to exist on computer servers whether or not there are active avatars within the virtual space (Bell 2008). This discrimination between VR and VW, however, is dissolving. VR experiences can be created to exist for days, and some users have been known to wear headsets for extended periods of time. Additionally, more and more VR experiences are being designed to be for game play, socialization, or mental relaxation. The IEEE

VR 2020 online conference and the Educators in VR International Summit 2020 offered participants opportunities to experience conference presentations in virtual rooms as avatars while interacting with presenters and conference attendees (see Sect. 2.5 for more information).

Relevant to defining VWs, Correia et al. (2016) conducted a meta-analysis on the potential of using virtual worlds for learning and training (p. 407), while Mann et al. (2018) proposed a different definition of XR with variations of real and synthetic applications that make up their Multimediated Reality Continuum (p. 12), for another term. According to Mann et al. (2018, abstract), "As a new field of study, All Reality is multidisciplinary. We must consider not just the user, but also how the technology affects others, e.g. how its physical appearance affects social situations, and how sensor-based reality (e.g. wearable and implantable cameras in the smart city) affects privacy, security, and trust. All Reality includes Virtual Reality (VR), Augmented Reality (AR), X-Reality (XR), X-Y Reality (XYR), and Mixed, Mediated, etc. realities (MR)."

CVEs—Collaborative virtual environments are communication systems in which multiple interactants share the same three-dimensional digital space despite occupying remote physical locations (Yee and Bailenson 2006).

Embodiment—Embodiment is defined by Lindgren and Johnson-Glenberg (2013) as the enactment of knowledge and concepts through the activity of our bodies within an MR (mixed reality) and physical environment (p. 445). Embodiment can also be experienced as a suspension of disbelief while using avatars (digital individual representations) in a fully online virtual world. In fact, embodiment can be experienced as group phenomena that may lead to the development of communities of practice (CoP). Truman (2014) studied the relationship of embodiment in collaborative virtual environments (CVEs) and its reflexive properties of the primary avatar (learner/user) related to the theoretical framework of transdisciplinarity (p. 59). Attachment to avatars resulting in embodiment was not always found, suggesting that some individuals may be incapable of embodied experience (Truman 2014: 232).

The deeply immersive nature of some forms of XR has had a powerful effect in studies of multifaceted empathy. Manipulating the full environment around a learner, including sight, sound, smell, taste, pressure, heat, and texture promises to be significant in impact. "VR feels real, and its effects on us resemble the effects of real experiences. Consequently, a VR experience is often better understood *not as a media experience, but as an actual experience,* with the attendant results for our behavior" (Bailenson 2018: 46). Studies have shown that learners can quickly adopt an avatar as a personal representation of their own bodies; this effect is known as body transfer. When learners accept a digital object as a real object, the Proteus effect (Yee and Bailenson 2007) has been achieved, and manipulations of the digital object are biochemically accepted as real (Fox et al. 2016). Each of these defined forms of technology, MR, AR, VW, and XR, provides learners with real experiences.

1.2   Key Trends

XR is becoming ubiquitous across society in domains such as entertainment, healthcare, government, military, education, and industry training for manufacturing and automation. Three key trends that are influencing the adoption of XR include:

1. The entertainment sector's leadership in promoting societal acceptance of immersive applications drive down the consumer costs for equipment used in virtual, mixed, and augmented reality. Examples include Nomadic that offers team-based, immersive gaming in a physical environment (Nomadic 2019) and The Void, created by the Walt Disney Company (Disney, 2019). When consumers have access to trying XR applications in a rich-media context without having to buy first, it is predicted that accelerated adoption will occur for campus and home-based uses.
2. Pervasive use of XR is leading to the creation of new forms of partnerships and the potential for mass collaboration. More sustainable open-source software communities are examples of collaborations that build ecosystems for virtual world platform development. University involvement in developing the XR ecosystem has been largely ad hoc. Potentially, if universities coordinate to design and develop communities of practice using XR, standards will develop faster for interoperability, building a robust ecosystem. The Advanced Distributed Learning (ADL) Initiative (2019), an organization of the US Department of Defense (DoD), provides analogous support for interoperability, reusability, open standards, and architecture. The most recent developments include the Total Learning Architecture (TLA) that includes the Competency Management System (CaSS), learner modeling for analytics development, e-learning standards, and the Sharable Content Object Reference Model (SCORM). DoD innovations funded and supported by ADL must meet security requirements.
3. The combination of sensors (Internet of Things), 2D virtual (web) environments, 3D immersive environments, and virtual experiences is a global phenomenon representing massive, rapid change that will impact society. When Industry 4.0 or the Fourth Industrial Revolution (4IR) combines with the third wave of the Internet (Internet of Everything), XR will be poised for ubiquitous use across society (Dindar et al. 2009: 34). The World Economic Forum identified 12 core areas that make up 4IR. These are "big data, artificial intelligence (AI), and internet of things (IoT), virtual and augmented realities, additive manufacturing, blockchain and distributed ledger technology, advanced materials and nanomaterials, energy capture, storage and transmission, new computing technologies, biotechnologies, geoengineering, neurotechnology, and space technologies" (Lieu et al. 2018: 2753).

Taking these trends into consideration, the following section explores XR learner activities with respect to STEM content and instructional design.

2 Use of XR in 2026

During a 2018 X-FILEs Workshop, educators, researchers, and professionals gathered to discuss and hypothesize about the future of XR within STEM higher education. Ideas generated from this workshop formed the basis for this chapter.

2.1 Content Presentation

Faculty and instructors typically think about instructional content to present, chunk, and sequence in their courses. Incorporating XR for graded assignments requires departmental and ideally institutional support to make the most of purchasing and design decisions. Appropriate XR applications can provide the foundations for new types of learning environments and experiences. XR can also bring users together creating new communities of inquiry and practice. With adequate support, XR can help users transcend the boundaries of classroom and web-based instruction. XR applications are discussed in this section in terms of the context of faculty demands and institutional resources based upon variables of size, staffing, need to scale, and emphasis on research focus. Taxonomies, toolkits, and means to obtain data valuable for creating baseline analytics using XR all support scholarship and research regardless of institution type.

2.1.1 Opportunities

XR presents an opportunity for an entirely new approach to learning design that can leverage the principles of transmedia learning. Transmedia learning provides a guide for educational experiences in XR, primarily because it offers the learner a real or simulated world with extra layers of overlaid information. As the learner is often at the center of these XR experiences, transmedia learning builds upon creating stories across experiences and devices where the learner is the hero possessing self-determination and self-regulation. The use of game-based learning, serious games, and gamification is included under Raybourn's (2018) transmedia learning framework, which provides heuristics for designing and personalizing transmedia learning. According to Raybourn (2018), "The transmedia learning framework (TLF) is meant to provide ways to think outside the box about learning, or what we normally consider education. The TLF is a way to employ new media in a way that is going to augment your learning experience, to include engaging yourself at the neurological level" (Raybourn 2018: 1).

2.1.2 Challenges

Authoring 3D content has become easier, but is not easy enough for fast production by most users and especially faculty. Library leadership is needed to track 3D open educational resources and provide authority for managing data into existing information schema and standards. Staff support for matchmaking of needs and content will require significant commitments of time to experiment and play with XR possibilities. Partnering with innovators on campus already doing XR can result in valuable informal professional development offerings that include panel discussions, hands-on demos, and tracking of effective uses.

2.1.3 Implementation Strategies

Instructors should be very clear about why they are selecting XR learner activities as opposed to other modes of teaching and learning. Apparent gains in learning, or increased learner preference, by utilizing XR can attributed to the novelty effect (Clark 1985) if compared to non-robust, non-immersive learning choices. Aldrich (2009) suggests that higher interactive virtual environments (HIVE) do work as learning activities because immersive activities stimulate emotional involvement, which is necessary for learning to move from working memory to long-term memory.

2.1.4 Research Questions

Future research relevant to content preparation in XR should address the following:

- What are the rights and ethical decisions that need to be made within organizations dedicated to XR development?
- How can common interface usage be created to apply to the design of XR experiences? For example, Control-C means copy in multiple software programs.
- What heuristics can be developed to increase understanding and outcomes of which XR application is most appropriate for various learning situations?
- How can XR facilitate collaboration on distributed team-based projects that are part of real systems used in society?
- How can incorporation of XR augment tutoring, counseling, help desks, or other virtual interactions?

2.2 Interactions and Communications

2.2.1 Opportunities

Some of the early evidence that interactions with XR applications are effective comes from corporate training (Maxwell and McLennan 2012), which suggests that adult learners experience enhanced transfer, possibly due to situated learning and because such training often has an immediate real-world application. This point should not be lost; Adult learners appear to be the first to have larger-scale learning success when focusing on the application of learning.

Collaborative virtual environments in the workplace allow for co-design and co-development to embed scenarios or situational problem-solving questions into immersive 3D environments. Microsoft's HoloLens 2 is innovating in the workplace environment by bringing what they call "instinctual interaction" into XR use (Upload 2019). The company is currently expanding into the workforce market, hoping to tap up to two billion frontline workers with HoloLens (Weise 2019).

Researchers are still in the nascent stages of discovering how the power of XR can be optimized. While studies by Yee and Bailison and others have examined the role that XR activities can play in behavioral skills like eliminating negative stereotyping, fostering empathy, and increasing scientific inquiry thinking processes, research on the concept of mirror neurons as the conduit for behavioral skills is maturing past conceptual to actual stages (Rizzolatti and Craighero 2004; Fox et al. 2016). Early data is pointing to the fact that the brain believes that XR activities really happen and as such, the internal biochemical changes of learning are the same as the real experience (Bailenson 2018). For example, XR can provide the benefits of field trips without requiring travel. Further XR ideas include: manipulating space and time; such as looking inside of a supernova; using expensive lab equipment at little cost; and scenarios as yet unimagined that are too dangerous or difficult. Thus XR can bring a new range of experiences to the learners of tomorrow.

For health, XR can also have a positive impact in the form of aiding cancer treatments (Chirico et al. 2016). The arrival of the COVID-19 pandemic and the closing of contact sports are causing a rise in the use of eSports. If, following the mirror neuron research from Rizzolatti and Craighero (2004), the mind believes that what the eyes see is actually happening, it is logical to assume that engaging in eSports could have a positive impact on physical health. During eSports, the body is being flooded with the same biochemicals as traditional athletes. Increased physical health could be measured with two blood pressure tests, one at each end of an academic semester for eSports athletes.

Research findings on empathy studies using XR are mixed at this time. Herrera et al. (2018) found that there are positive long-term impacts via VR experience on the topic of homelessness. However, a VR experience with color-blindness resulted in no short-term empathy increases. In some cases, exposure to VR decreases empathy. Companies like Equal Reality (https://equalreality.com/) are using virtual reality for diversity training as a way to impact perspectives on privilege and power.

2.2.2 Challenges

XR platforms also have significant accessibility concerns that need to be addressed with both hardware and software. Many experiences require full 360° movement and motion controller input and do not have any accessibility features built in to allow for a wider variety of users. There are a variety of ways that this affects users with a range of abilities (Ryan 2019). Opportunities to create compelling social XR applications are exciting but must not come at the cost of alienating users.

When we introduce multiple users into an XR experience, we introduce the possibility for a whole host of issues that are not encountered in single user educational experiences. We encounter significant issues related to safety, accessibility, and privacy that must be accounted for very early on in the design process of any multi-user educational experience (Reilly et al. 2014).

When using existing social virtual reality experiences, female users reported struggling "in a social sense, dealing with strangers and encountering behaviors that they equated to real life scenarios where they had experienced harassing behavior. In the social VR worlds they visited, they encountered flirting, a lack of respect for personal boundaries, socially undesirable behavior and in the end, most were not interested in using these platforms to meet strangers" (Outlaw and Duckles 2017). The social problems we deal with in location-based educational environments follow us into XR. Outlaw and Duckle's study of women in social VR experiences highlights the issues of personal and social safety faced by those who experience a lack of social safety outside of VR. In VWs, the phenomenon known as griefing could negatively impact digital systems in the form of attacks that reduce digital commerce (Bakioglu 2009).

2.2.3 Implementation Strategies

Educators should be looking for ways to leverage existing tools to create meaningful interactions using XR technologies while continuing to explore new opportunities that have not yet been developed. "Remote real-time collaboration with socio-technical systems and dialogue tools aimed at promoting collaborative learning and deepening the space of debate and producing epistemic interactions is in the interest of designers, engineers and educators around the globe. This calls for enabling more platforms for real-time collaboration between teams and networks" (Wendrich et al. 2016). XR technologies will also provide new forms of leadership engagement between students, faculty, and institutions.

Using commercially available applications, like Altspace or Bigscreen, educators already have the opportunity to present more traditional digital content (video, photos, presentations, etc.) to students in an XR environment. Utilizing these types of systems allows for a low barrier to entry that requires minimal customization or onboarding. Educators can quickly adapt existing content to be used.

For universities that already have existing XR equipment, labs, and studios, there are great opportunities to invite students deeper into the design process and to create collaborations with other universities so students can interact with other students who might have similar access. It should be a top priority for institutions with these resources to leverage them, in order for students to connect with their peers at other universities in XR experiences.

2.2.4 Research Questions

Future research regarding interactions and discussions using XR should address the following:

- How can XR improve peer-to-peer learning?
- How do people construct their own virtual identity?
- How do XR technologies differ in their effectiveness for fostering meaningful discussions?
- What factors cause learners from different backgrounds to be inhibited in multi-user XR experiences?

2.3 Learner Activities with XR

Prior to 1910, STEM classes were taught primarily with direct instruction (Olson and Loucks-Horsley 2000). This started to change with John Dewey's emphasis on active learning activities that occurred outdoors, outside the science classroom. Over time, however, STEM teachers wanted activities that would be active in cognitive load but could also be done indoors to eliminate problems of weather, distance, and availability (Brown 2003). The space-era Commission on Science Education of 1964 has strongly influenced STEM instruction for the past half-century by requiring the inclusion of experiential learning approaches, known as laboratory courses (labs), in addition to rote learning of the scientific method (Livermore 1964). Virtual labs provide equal learning gains when compared to in-person labs (Faulconer and Gruss 2018). The emphasis in XR, however, is less on content knowledge and more on inquiry scientific thinking processes and behaviors and then collaborating and communicating. Consequently, VR, VW, AR, and all forms of XR have a history of providing the valid and immersive active learning activities required within STEM domains (Nelson and Ketelhut 2007).

2.3.1 Opportunities

Designing with XR requires a fundamental shift to a Human-Centered Design philosophy that involves putting human needs, capabilities, and behavior first (Jerald 2018: 15). XR provides the opportunity to experience just-in-time *immersive*, experiential learning that uses concrete yet exploratory experiences involving senses that result in lasting memories. Here we discuss opportunities for social applications with XR.

XR learner activities are usually created for individual use, which may or may not need to be simultaneously experienced as a class together at the same time or place with the instructor. Activities can be designed into instruction with VR headsets, high-resolution screens, smartphones, or other solo technological devices for use inside and outside of the classroom. Feedback is also often individualized (Lynch and Ghergulescu 2017). Multiple characteristics contribute to the growing ubiquity of XR.

STEM XR learner activities often have these characteristics:

- Decreased danger to learners, increased respect for the environment, and replicability—chemicals and radiation are virtual and can be used indefinitely (Faulconer and Gruss 2018). Inside VR, "mistakes are free" (Bailenson 2018: 24). In virtual reality, chemicals can be mixed in the wrong order and cleaned up just by pressing a recycle button (Faulconer and Gruss 2018). Surgical emergencies can be practiced with no ill effects upon real patients (Health Scholars n.d.).
- No limits to time or space—learners can control time within a biological system (Clark 2009), or learners can go to the International Space Station. There can be full instructor control of the virtual experience, important when engaging in dangerous or trauma-replicating activities (Bailenson 2018). XR has the power to make difficult science concepts to learn easier by "making the unseen seen" (Potkonjak et al. 2016: 4).
- Increased respect for learner ethics and accessibility needs—virtual dissections reduce need for organisms, and text and sound can be added as layers of additional information, that information can be more thorough upon demand, and XR resources paused, reset, and can be available anytime (Lynch and Ghergulescu 2017).
- Decreased cost due to low maintenance and the ability to easily replicate the experience for more learners (Faulconer and Gruss 2018).

Some published examples of the use in XR learner activities within STEM include:

- Genome Island within the VW Second Life provides asynchronous learning experiences where time can be accelerated and multiple generations of organisms can be studied within a few minutes (Clark 2009).
- Labster provides virtual labs that go beyond simple bench work and include narratives that show the role of collecting samples, safety protocols, and analyzing results (Pate 2020).
- Arizona State University partnered with Smart Sparrow to make "immersive, interactive virtual field trips" (Mead et al. 2019: 2) that allow for instantaneous movement around a constructed paleoenvironment.
- Functional analysis of historic architectural structures to understand engineering innovation of buildings. Notre Dame is an example (CBS Interactive Inc. 2019).

- Learners can construct their own representation of their knowledge. The app Orb uses AR to allow students to create simple 3D objects that appear in real space (Donally 2018).
- Restivo et al. (2014) found that the use of AR in teaching direct current circuits allowed learners to have the ability to overlay real-world and real-time lab experiences.

2.3.2 Challenges

The technological challenges of the use of XR in learner activities should be acknowledged. There will always be first-use hesitation from learners, and they may tend to prefer more traditional activities until the newer technological interface becomes intuitive (Waldner et al. 2006). VR head-mounted displays can be burdensome with tethering cables and tight-fitting headsets, and if technology is shared or passed from learner to learner, there are the added challenges of the sharing of germs and body moisture. Further developments already coming in VR equipment include the use of temperature, pressure, and scent which can potentially disturb the learner. Bailenson aptly identifies the VR dangers of poor behavioral modeling, simulator sickness, eyestrain, and reality blurring (2018). VW can be addictive and can contain significant non-educational activities not suitable for young learners. AR can cause distractions if used walking down the street or while driving.

XR learner activities should be designed to increase accessibility to all learners by utilizing the XR strengths of multi-layered information display. However, some learners will be held back from full XR activity by visual, physical, and social abilities such as stroke, vertigo, epilepsy, or age-related reaction time. It should also be noted that the encompassing nature of VR headsets might create some discomfort or danger for any learners as they can no longer fully see and control their body and body space.

Keeping young learners focused while using technology can be a challenge. Tallyn et al. (2005) found that AR could bring media-based learning to life when used in concert with paper-based learning as an instructional theme through a learning experience. Learners focused on working through a worksheet or workbook could reap the benefits of AR while not going off-track with their learning.

2.3.3 Implementation Strategies

Combining VR applications such as virtual worlds with other XR applications can serve as a multiplier effect for research where communities of practice span across international boundaries. User support communities are robust in the virtual worlds of Second Life and OpenSimulator where institutional representation often occurs at the grassroots level. Immersive conferences such as the Virtual Worlds Best Practices in Education and the OpenSimulator Community Conference enable sharing of R&D across XR applications. Network members in organizations such as the Immersive Learning Research Network and Educators in VR overlap the disciplines of education, technology, and industry in order to further the research-based effective application of XR technologies.

Examining how augmented reality and virtual reality social systems differ, how they may become blended, and the necessary design considerations, we can see how as these technologies converge there will be new possibilities for exploration and collaboration (Miller et al. 2019). Already there are social XR platforms, like AltspaceVR and Bigscreen, that allow users with access to a variety of XR devices to share virtual spaces and content from their own devices with each other.

In juxtaposition to the experiences of users in 3D worlds using 2D screen, or those using video-based telepresence systems, XR social platforms allow for users to feel as if their "virtual self is experienced as the actual self" (Aymerich-Franch et al. 2012). This sense of social presence (Oh et al. 2018) offers new areas of exploration for users of this technology. We can now consider not just adapting existing educational content, but conceiving of entirely new ways of educating that leverage an embodied experience and allow for a huge range of augmentation.

The affordances of XR environments include opportunities for geographically dispersed learners to learn in an environment similar to their traditional classrooms without forfeiting the ability to learn at their own pace and in their own time zone (Olasoji and Henderson-Begg 2010). Computer-supported collaborative work (CSCW) or team-based projects involve looking at the individual and collaborative potentials based on individual and shared contributions. Ward and Sonneborn (2009) note that methods of assessing creative problem-solving in groups will need to include measures of how people personalize their learner group contributions and the effect that such individualized collaboration has on the quantity and quality of ideas produced.

2.3.4 Research Questions

Future research in learner activities using XR should include:

- What learner activities in XR are active learning approaches, as opposed to passive approaches?
- How are learner activities in XR fostering critical STEM skills including inquiry, empathy, collaboration, and communication?
- How do XR learning activities impact human physiological and anatomical structures of the brain, such as for improving self-regulated learning with biofeedback from usage of wearables tracked by avatars?
- How can XR learning activities foster play as a foundation for learning?

Since her earliest school years, Andi has been going on virtual field trips that utilize basic VR headsets in the classroom and immersive rooms at museums. Her initial experiences involved little interaction and were more look-and-see experiences, such as taking a field trip to the Great Wall of China while using a VR headset. Gradually, Andi's teachers added interactive experiences during which Andi learned to navigate controllers to draw, assemble, and bounce virtual objects in small group activities with her classmates. By middle school, Andi was a confident presenter in virtual reality platforms and often played social VR games outside of school. In 6th grade, Andi wrote a report on her virtual visit to Mars, ending with her resolution to become an astronaut.

In high school, Andi enjoyed participating in eSports and played on an all-female team with the potential to win $30,000 in college scholarships. She also engaged in weekend virtual science workshops on biofuel creation sponsored by corporations looking to increase students' scientific collaboration skills of observation, cause-and-effect monitoring, and communication of results. Andi regularly joined virtual check-ins with scientists at international museums and connected virtually with astronauts at the National Aeronautics and Space Administration and the International Space Station.

2.4 Assessment

2.4.1 Opportunities

The immersive nature of XR has the potential to provide engaging assessment experiences that represent real-world activities more accurately than traditional paper and pencil measures. Still, assessment of learning in virtual environments typically employs the use of external instruments that are administered outside of the virtual experience. This review examined the literature related to the types of assessments used in the context of XR, with a focus on methods for assessing skills, evaluating work products and performance, and analyzing log files.

Conventional assessments, such as pre and post, multiple-choice items, short answer items, and open-ended questions, are frequently used to assess the effectiveness of learning in a virtual environment. Ketelhut et al. (2006) created a pre-post, self-designed content test to assess knowledge of science inquiry and process skills during an investigation of a science curriculum, implemented through a virtual environment called River City. Labster's HMD and desktop VR lab experiments include embedded multiple-choice questions and a point system to track students' scores as they progress through experiments. Metrics, such as assessing the number of questions asked in a lab experience within and without a pre-lab XR experience, offer additional possibilities for collecting performance measures.

In contrast, evaluating learner-created work products developed directly within immersive virtual environments provides opportunities for assessing authentic learner performance. Rose (1995) leveraged the psychological theories of information processing and constructivism to identify specific approaches for measuring learning in VR that included performance tasks, such as world building, problem-solving, and the evaluation of the quality of final products. Olasoji and Henderson-Begg (2010) studied a Second Life course that required learners to produce a summative assessment containing a scientifically accurate depiction of a biological molecule or bioinformatic concept. Work products such as these can be evaluated to produce data concerning the learner's performance through a rubric, a scoring guide, or an automated scoring procedure (Mislevy et al. 2017).

The Virtual Performance Assessment (VPA) Project, created by the Harvard Graduate School of Education, is an open-ended 3D immersive virtual environment designed for performance assessments of science inquiry skills in multiple virtual scenarios. Students engage in authentic inquiry activities and solve scientific problems by navigating around the virtual environment as avatars, making observations, interacting with non-player characters (NPCs), gathering data, and conducting laboratory experiments (Baker and Clarke-Midura 2013). VPA enables the automated and non-intrusive collection of process data (event logs or logged actions and behaviors) and product data (students' final claims), facilitating the capture and assessment of science inquiry in situ (Jiang et al. 2015). These examples represent advancements toward relevant methods for assessing learning within virtual settings.

A frontier of educational assessment is the development of automated methods of evaluating log files, through which cognitively meaningful patterns and features of work are detected and characterized as observations (Mislevy et al. 2017). The dynamic nature of the computer system allows recording of learner interactions and data gathering in the background as the learner moves through an XR experience (Rose 1995). Log file analysis of this type of data is often referred to as stealth assessment, which has been used to measure problem-solving and spatial skills (Shute et al. 2015), assess causal reasoning in the World of Goo (Shute and Kim 2011), and assess systems thinking in Taig Park (Shute et al. 2010). An advantage of stealth assessments is that data can be collected without disrupting learner flow within the XR experience (Shute et al. 2015).

2.4.2   Challenges

Assessments in XR present several challenges that can impact the integrity of the inferences that can be made from results. Examples include interference due to headset discomfort, lack of familiarity with navigating the virtual world, and gaps in alignment between the assessment method and the content being assessed. Construct-irrelevant variance caused by distractions in the virtual environment can negatively influence learner behavior. In multi-user environments, issues around trust, security, and identity can arise, given that users can create multiple accounts and avatars (Warburton 2009).

In the context of stealth assessments, the complexity of log file coding schemes can make analysis difficult, while a lack of aligned external validation instruments can pose additional reliability challenges (Wang et al. 2015). According to Wang et al., "If a researcher wants to create stealth assessments within an existing commercial game, the first step is to make sure that either the coding in the log files is simple enough to understand, or the coding scheme is available from the game developer so that changes can be made to the information that is being captured" (2015: 5). It is important to note that not every logged interaction or tracked gaze represents evidence of learning. Validating accurate scoring systems for virtual activities that have the ability to predict performance in real-world settings presents a formidable challenge for XR assessment developers.

2.4.3 Implementation Strategies

Implementing assessments that have the capacity to elicit evidence of what learners know and can do requires close alignment between learner interactions and the expected outcomes (Code et al. 2012). By leveraging the power of evidence-centered design (ECD) (Mislevy and Haertel 2006), assessment developers can create an evidentiary assessment argument supported by a conceptual assessment framework that includes task specifications, evaluation procedures, and measurement models. In a virtual environment, provisions for accommodations and the use of the principles of universal design for learning (UDL) are critical for ensuring accessibility and equity during assessment implementation.

2.4.4 Research Questions

Overall, the development of reliable assessment methods within XR is still in the early stages. Future research is needed to examine the following research questions:

- What indicators will inform whether contextual learning using XR is connected to curriculum alignment?
- What are the most effective and useful means of tracking competency for various skills?
- What is the efficacy of novel forms of assessment within XR?
- How can learning and performance be accurately assessed in XR environments?
- What methods of assessment are possible in XR environments?

2.5 Co-curricular Activities

Here we explore the experience of learners in the context of academic endeavors beyond classroom interactions that include socialization, employment, and responsibilities that provide opportunities, challenges, and potential for future research

To mitigate safety risks during Andi's first XR experiences, her teachers limited the viewing to under 2 min. Teacher assistants served as spotters, and the space for engaging with the hardware was cleared of hard surfaces. Alternative experiences were set up in 2D for any learner.

One of Andi's teachers wisely included safety precautions in her XR choices and picked a platform where the students were not tracked and did not need to log-in with authentic credentials. When the learner's session ends, the entire session is deleted from the host's servers. At the high school computer lab, the instructional technology specialists set up all of the VR devices to be controlled and monitored from a main station and ensured that devices could not be accessed outside of the school network without specific teacher permission. Andi's instructors selected learning experiences where teleportation was used rather than abrupt motion, to reduce incidences of motion sickness.

Routinely, XR devices were disinfected daily during the school year. Learners were taught to wipe down the headset and controllers before using them. All parents were informed of the planned instructional use of XR through the year, and opt-in signatures were requested.

Andi's early choices within XR environments were limited to looking at a spot, clicking on something, a small range of "right click" alternative choices, and click and drag. As she gained confidence, she navigated multiple screens (e.g., in XR and in browser) at the same time, learned fly commands, combined virtual objects, and panned around 3D depictions. Due to products like Unity and other 3D content creators, Andi was able to create unique XR experiences with new textures and objects. Her instructors used her content creations within XR as evidence of learning, rather than relying on traditional forms of assessment outside of XR.

using XR. Testing and adoption of commercial XR applications by learners enables instructors to harness new data that is useful for providing analytics of learner performance while supporting motivation to engage in STEM. The National Academies of Sciences, Engineering, and Medicine (2016: 95) reported that the use of co-curricular programming can affirm students' self-perceptions of competence to mitigate impacts of a stigmatizing STEM academic culture.

2.5.1 Opportunities

The pervasive use of XR is leading to the creation of new forms of partnerships and the potential for mass collaboration. More sustainable open-source software communities are examples of collaborations that build ecosystems for virtual world platform development. Further expansion of these collaborations is recommended to advance the field of XR.

Obtaining value from the use of XR applications requires new interdisciplinary collaboration across departments, campuses, and partnerships with industry to manage complexity in pursuit of improved learning outcomes. Virginia Tech created a report for envisioning the campus of the future where organization and reorganization will be required among stakeholders including the greater community to fuse intellectual and co-curricular life as Human-Centered Smart Environments (HCSE). Figure 1 illustrates a vision for technology-enabled, interdisciplinary participation in societal needs that create new forms of learning opportunities. The emphasis in this diagram is that collaboration from multiple sources is necessary in human-centered environments. This collaboration needs to be made and remade in successive cycles in order for the positive impacts to be effective. For example, maker-spaces and research efforts on college campuses need to collaborate with industry to make the results widely available.

Virginia Tech's HCSE plans include Living Laboratories that are "... experiential environments wherein students and researchers can engage in learning, discovery, and innovation through interactions with real-world systems and communities. Living labs incorporate 'smart' autonomous systems that range from furniture and wearable technology to buildings and the surrounding landscape. These systems will assist humans in multiple ways that include supporting wellness, collaboration, solitude, inspiration, and disruption" (Hundley 2016: 11). The Virginia Tech HCSE

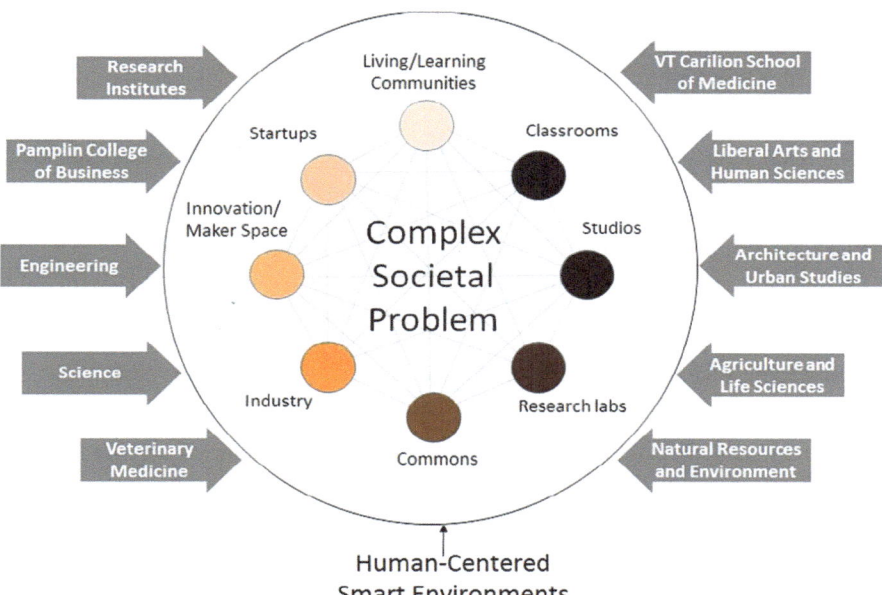

Fig. 1 Hundley (2016: 8) integrated innovation hubs

model represents the framework for a type of immersive, digital learning commons that can incorporate existing instructional and information software systems, designed for the most common XR applications used across the institution.

2.5.2 Challenges

The field of XR is in need of the creation of standards and norms for implementing technologies in educational settings, which could provide support for and accelerate their adoption by higher education institutions. XR applications currently do not have universal standards that enable interoperability and seamless integration into existing software. Professional associations such as the IEEE's IC Industry Consortium on Learning Engineering (ICICLE) (https://www.ieeeicicle.org/) XR working groups (https://www.ieeeicicle.org/xreality-sig) are investigating how to make the user experience more seamless and transparent. Potentially, if universities coordinate to design and develop communities of practice using XR, standards will develop faster for interoperability, building a robust ecosystem. The pervasive use of XR is leading to the creation of new forms of partnerships and the potential for mass collaboration. More sustainable open-source software communities are examples of collaborations that build ecosystems for virtual world platform development. Further expansion of these collaborations is recommended to advance the field of XR.

2.5.3 Implementation Strategies

Recommendations for higher education institutions and faculty seeking to advance the use of XR in co-curricular instruction include:

- Promoting interdisciplinary connections (ways to encourage and support different ways/lenses of looking at issues or problems) such as learner-produced content and authoring. Examples of this approach include student hackathon events and leadership engagement with clubs and organizations.
- Expanding access to virtual internships and field trips.
- Broadening collaborations across academics, industry, and community (making connections) including collaborating with the military for bootstrapping research and development.
- Increasing access to industry and government facilities and tools using XR.
- Creating standards and norms to support implementation.

During the COVID-19 pandemic, educational institutions of all sizes implemented XR as a social connection strategy. Platforms such as Discord, Mozilla Hubs, AltspaceVR, and VirBELA which were previously used only for co-curricular meetings and game spaces became more popular. The power of these platforms

As a college senior, Andi has experienced several courses in virtual worlds and designed her dorm room using the Walmart AR app to make furniture choices. Training for her current job has included virtual simulations that address complex problems through global collaborations and negotiations among individuals and large groups in virtual settings. She regularly attends virtual conferences that connect her with thousands of attendees in virtual spaces, such as Altspace, Mozilla Hubs, and VirBELA. Utilizing industry and media contacts, Andi can include any of her family and friends in XR activities because she can find XR resources that work for a range of ages and abilities. Many XR experiences have text, as well as icon labels so that Andi can engage with learners that need language-flexible choices. Andi's father often engages in VR training before he visits a new building site, while her mother uses XR training to practice empathy skills with her fellow emergency medical technicians. For Andi and her family, XR has become a common part of daily life.

was leveraged to conduct virtual global conferences, such as the Immersive Learning Research Network 2020 Conference, the IEEE VR 2020 Conference, and the Educators in VR 2020 Conferences. These events engaged thousands of students, educators, and industry practitioners from around the world in virtual convenings that demonstrated how XR resources can be used for the purpose of sharing knowledge and professional networking, when traditional conference formats are impossible.

2.5.4 Research Questions

Recommendations for areas of future research related to co-curricular XR include the following research questions:

- What are the mental, physical, and neurological effects of using XR technology on the learner for various durations (including long-term)?
- Which XR learning activities increase time on learning tasks?
- How can the effective use of XR technologies improve collaboration practices for transdisciplinary scholarship and research?
- How does XR support improvement among individuals and teams, especially at a distance?
- How do institutions leverage XR to create communities of practice to promote cross-institutional, cross-industry, and cross-domain collaboration of R&D?

3 Conclusions

As posited in the chapter introduction, the influence of the entertainment sector, new collaborations between technology and business, and the ubiquity of the Internet of Things all indicate that the required technology for XR might already be in American homes, classrooms, and workplaces in the form of smartphones, computers, or game systems. "VR alone is expected to reach $60 billion in 2020" (Bailenson 2018: 9), and AR is expected to reach $60 billion in 2020 (Porter and Heppelmann 2017). With the COVID-19 pandemic, institutions of all sizes are considering the incorporation of XR experiences.

In conclusion, while there are many opportunities for students and faculty who are able to leverage the benefits of XR to enhance coursework, a variety of obstacles may hinder scaling. Optimizing the use of XR in higher education requires the support and resources of an interdisciplinary community of committed professionals from education, government, and industry who will work together with researchers to overcome the existing challenges that limit adoption. The development of common standards could advance this effort. There is also a need to invest in future research regarding the implementation and the assessment of the effectiveness of XR on learning. Technological and practical solutions are possible through the collaboration of experts and the financial investment for research in this field.

References

Advanced Distributed Learning Initiative (2019) About. https://www.adlnet.gov/about

Aldrich C (2009) Learning online with games, simulations, and virtual worlds. Jossey-Bass, San Francisco

Aymerich-Franch L, Karutz C, Bailenson JN (2012) Effects of facial and voice similarity on presence in a public speaking virtual environment. In: Proceedings of the international society for presence research annual conference, Philadelphia

Bailenson J (2018) Experience on demand: what virtual reality is, how it works, and what it can do. W. W. Norton & Company, New York

Baker RSJD, Clarke-Midura J (2013) Predicting successful inquiry learning in a virtual performance assessment for science. In: Carberry S, Weibelzahl S, Micarelli A, Semeraro G (eds) User modeling, adaptation, and personalization. UMAP 2013. Lecture notes in computer science, vol 7899. Springer, Berlin, Heidelberg

Bakioglu BS (2009) Spectacular interventions of second life: goon culture, griefing, and disruption in virtual spaces. J Virtual Worlds Res 1(3)

Bell MW (2008) Toward a definition of "virtual worlds". J Virtual Worlds Res 1(1)

Brown F (2003) Inquiry learning: teaching for conceptual change in EE. Green Teach 71:31–33

CBS Interactive, Inc. (2019, 17 April) Video game "Assassin's Creed" could play a role in Notre Dame Cathedral's restoration. https://www.cbsnews.com/news/notre-dame-cathedral-fire-video-game-assassins-creed-could-help-in-its-restoration/

Chirico A, Lucidi F, De Laurentiis M, Milanese C, Napoli A, Giordano A (2016) Virtual reality in health system: beyond entertainment. A mini-review on the efficacy of VR during cancer treatment. J Cell Physiol 231(2):275–287

Clark RE (1985) Evidence for confounding in computer-based instruction studies: analyzing the meta-analyses. ECTJ 33(4):249–262

Clark MA (2009) Genome Island: a virtual science environment in Second Life. Innov J Online Educ 5(6)

Code J, Clarke-Midura J, Zap N, Dede C (2012) Virtual performance assessment in immersive virtual environments. In: Wang H (ed) Interactivity in e-learning: cases and frameworks. IGI Publishing, New York, pp 230–252

Correia A, Fonseca B, Paredes H, Martins P, Morgado L (2016) Computer-simulated 3D virtual environments in collaborative learning and training: meta-review, refinement, and roadmap. In: Sivan Y (ed) Handbook on 3D3C platforms. Springer International Publishing, Cham, pp 403–440. https://doi.org/10.1007/978-3-319-22041-3_15

Dindar N, Balkesen Ç, Kromwijk K, Tatbul N (2009) Event processing support for cross-reality environments. IEEE Pervasive Comput 8(3):34. https://ieeexplore.ieee.org/stamp/stamp.jsp?arnumber=5165558

Donally J (2018) Learning transported. Augmented, virtual, and mixed reality for all classrooms. International Society for Technology in Education, Portland

Faulconer E, Gruss A (2018) A review to weigh the pros and cons of online, remote, and distance science laboratory experiences. Int Rev Res Open Distrib Learn 19(2). https://doi.org/10.19173/irrodl.v19i2.3386

Fox NA, Bakermans-Kranenburg MJ, Yoo KH, Bowman LC, Cannon EN, Vanderwert RE et al (2016) Assessing human mirror activity with EEG Mu rhythm: a meta-analysis. Psychol Bull 142(3):291–313. https://doi.org/10.1037/bul0000031

Health Scholars (n.d.) Fire in the OR™ Virtual Reality Simulation I Medical Training For Surgical Fires. https://www.youtube.com/watch?v=10Ke4kDSpGM&feature=youtu.be

Herrera F, Bailenson J, Weisz E, Ogle E, Zaki J (2018) Building long-term empathy: a large-scale comparison of traditional and virtual reality perspective-taking. PLoS One 13(10):e0204494. https://doi.org/10.1371/journal.pone.0204494

Hundley M (2016) Envisioning the campus of the future. https://vtechworks.lib.vt.edu/bitstream/handle/10919/79636/BBenvisioning-the-campus-of-the-future-committee-report.pdf?sequence=1

Jerald J (2018) Human-centered VR design: five essentials every engineer needs to know. IEEE Comput Graph Appl 38(2):15–21

Jiang Y, Paquette L, Baker R, Clarke-Midura J (2015) Comparing novice and experienced students in virtual performance assessments. In: Proceedings of the 8th international conference on educational data mining, Madrid, 26–29 June

Ketelhut DJ, Dede C, Clarke J, Nelson B (2006) A multi-user virtual environment for building higher order inquiry skills in science. In: Annual conference of the American educational research association, San Francisco

Lieu TTB, Duc NH, Gleason NW, Hai DT, Tam ND (2018) Approaches in developing undergraduate IT engineering curriculum for the fourth industrial revolution in Malaysia and Vietnam

Lindgren R, Johnson-Glenberg M (2013) Emboldened by embodiment: six precepts for research on embodied learning and mixed reality. Educ Res 42(8):445–452

Livermore AH (1964) The process approach of the AAAS commission on science education. J Res Sci Teach 2(4):271–282. https://doi.org/10.1002/tea.3660020403

Lynch T, Ghergulescu I (2017, March) Review of virtual labs as the emerging technologies for teaching STEM subjects. In: INTED2017 Proc. 11th Int. Technol. Educ. Dev. Conf. 6-8 March Valencia Spain, pp 6082–6091

Mann S, Havens JC, Iorio J, Yuan Y, Furness T (2018) All reality: values, taxonomy, and continuum, for virtual, augmented, eXtended/MiXed (X), Mediated (X, Y), and multimediated reality/intelligence. http://wearcam.org/all.pdf

Maxwell D, McLennan K (2012) Case study: leveraging government and academic partnerships in MOSES (Military Open Simulator [Virtual World] Enterprise Strategy). In: Amiel T, Wilson B (eds) Proceedings of EdMedia 2012—world conference on educational media and technology.

Association for the Advancement of Computing in Education (AACE), Denver, pp 1604–1616. https://www.learntechlib.org/primary/p/40960/

Mead C, Buxner S, Bruce G, Taylor W, Semken S, Anbar AD (2019) Immersive, interactive virtual field trips promote science learning. J Geosci Educ 67(2):131–142

Miller MR, Jun H, Herrera F, Yu Villa J, Welch G, Bailenson J (2019) Social interaction in augmented reality. PLoS One 14(5):e0216290. https://doi.org/10.1371/journal.pone.0216290. https://journals.plos.org/plosone/article?id=10.1371/journal.pone.0216290

Mislevy RJ, Haertel GD (2006) Implications of evidence-centered design for educational testing. Educ Meas Issues Pract 25(4):6–20

Mislevy RJ, Haertel GD, Riconscente MM, Rutstein D, Ziker C (2017) Assessing model-based reasoning using evidence-centered design: a suite of research-based design patterns. Springer, New York. https://www.springer.com/us/book/9783319522456

National Academies of Sciences, Engineering, and Medicine (2016) Barriers and opportunities for 2-year and 4-year STEM degrees: systemic change to support students' diverse pathways. National Academies Press, Washington, DC

Nelson BC, Ketelhut DJ (2007) Scientific inquiry in educational multi-user virtual environments. Educ Psychol Rev 19(3):265–283. https://doi.org/10.1007/s10648-007-9048-1

Nomadic VR (2019) Nomadic—make believers. https://blurtheline.com/

Oh CS, Bailenson JN, Welch GF (2018) A systematic review of social presence: definition, antecedents, and implications. Front Robot AI 5:1–34. https://doi.org/10.3389/frobt.2018.00114

Olasoji R, Henderson-Begg S (2010) Summative assessment in second life: a case study. J Virtual Worlds Res 3(3). https://doi.org/10.4101/jvwr.v3i3.1460

Olson S, Loucks-Horsley S (eds) (2000) Inquiry and the national science education standards: a guide for teaching and learning. National Academies Press, Washington, DC

Outlaw J, Duckles B (2017) Why women don't like social virtual reality: a study of safety, usability, and self-expression in social VR. The Extended Mind, Portland. https://extendedmind.io/social-vr

PateAL (2020) Diverse avatars and inclusive narratives in virtual reality biology simulations. Emerg Learn Des J 7(1):Article 4. https://digitalcommons.montclair.edu/eldj/vol7/iss1/4

Porter ME, Heppelmann JE (2017, November 1) Why every organization needs an augmented reality strategy. Harv Bus Rev. https://hbr.org/2017/11/a-managers-guide-to-augmented-reality

Potkonjak V, Gardner M, Callaghan V, Mattila P, Guetl C, Petrović VM, Jovanović K (2016) Virtual laboratories for education in science, technology, and engineering: a review. Comput Educ 95:309–327. https://doi.org/10.1016/j.compedu.2016.02.002

Raybourn EM (2018) On demand learning for better scientific software: how to use resources & technology to optimize your productivity. [Blog post]. https://bssw.io/blog_posts/on-demand-learning-for-better-scientific-software-how-to-use-resources-technology-to-optimize-your-productivity

Reilly D, Salimian M, MacKay B, Mathiasen N, Edwards WK, Franz J (2014, June) SecSpace: prototyping usable privacy and security for mixed reality collaborative environments. In: Proceedings of the 2014 ACM SIGCHI symposium on engineering interactive computing systems, ACM, pp 273–282

Restivo T, Chouzal F, Rodrigues J, Menezes P, Bernardino Lopes J (2014) Augmented reality to improve STEM motivation. In: 2014 IEEE global engineering education conference (EDUCON), IEEE, Istanbul, pp 803–806. https://doi.org/10.1109/EDUCON.2014.6826187

Rizzolatti G, Craighero L (2004) The mirror-neuron system. Annu Rev Neurosci 27:169–192

Rose H (1995) Assessing learning in VR: towards developing a paradigm virtual reality roving vehicles (VRRV) project. Human Interface Laboratory

Ryan A (2019, January 27) Thoughts on accessibility issues with VR. https://ablegamers.org/thoughts-on-accessibility-and-vr/

Shute VJ, Kim YJ (2011) Does playing the World of Goo facilitate learning? In: Design research on learning and thinking in educational settings: enhancing intellectual growth and functioning, pp 359–387

Shute VJ, Masduki I, Donmez O (2010) Conceptual framework for modeling, assessing and supporting competencies within game environments. Technol Instr Cogn Learn 8(2):137–161

Shute VJ, Ventura M, Ke F (2015) The power of play: the effects of Portal 2 and Lumosity on cognitive and noncognitive skills. Comput Educ 80:58–67

Tallyn E, Frohlich D, Linketscher N, Signer B, Adams G (2005) Using paper to support collaboration in educational activities. In: Proceedings of the 2005 conference on computer support for collaborative learning learning 2005: the next 10 years!—CSCL '05. Association for Computational Linguistics, Taipei, pp 672–676. https://doi.org/10.3115/1149293.1149381

Truman BE (2014) Transformative interactions using embodied avatars in collaborative virtual environments: towards transdisciplinarity (3628698). Available from ProQuest Dissertations & Theses Global. http://ciret-transdisciplinarity.org/biblio/biblio_pdf/barbara_truman.pdf

Upload VR (2019) HoloLens 2 AR headset: on stage live demonstration—YouTube. https://www.youtube.com/watch?v=uIHPPtPBgHk&t=2s

Waldner M, Hauber J, Zauner J, Haller M, Billinghurst M (2006) Tangible tiles: design and evaluation of a tangible user interface in a collaborative tabletop setup. In: Proceedings of the 20th conference of the computer-human interaction special interest group (CHISIG) of Australia on Computer-human interaction: design: activities, artefacts and environments—OZCHI '06, ACM Press, Sydney, p 151. https://doi.org/10.1145/1228175.1228203

Wang L, Shute V, Moore GR (2015) Lessons learned and best practices of stealth assessment. Int J Gam Comput Mediat Simul 7:66–87. https://doi.org/10.4018/IJGCMS.2015100104

Warburton S (2009) Second life in higher education: assessing the potential for and the barriers to deploying virtual worlds in learning and teaching. Br J Educ Technol 40(3):414–426

Ward TB, Sonneborn MS (2009) Creative expression in virtual worlds: imitation, imagination, and individualized collaboration. Psychol Aesthet Creat Arts 3:211–221. https://doi.org/10.1037/a0016297

Weise K (2019) You're hired. Now wear this headset to learn the job. The New York Times [Newspaper]. https://www.nytimes.com/2019/07/10/business/microsoft-hololens-job-training.html

Wendrich RE, Chambers K, Al-Halabi W, Seibel EJ, Grevenstuk O, Ullman D, Hoffman HG (2016) Hybrid design tools in a social virtual reality using networked oculus rift: a feasibility study in remote real-time interaction. In: Volume 1B: 36th computers and information in engineering conference. https://doi.org/10.1115/detc2016-59956

Yee N, Bailenson JN (2006) Walk a mile in digital shoes: the impact of embodied perspective-taking on the reduction of negative stereotyping in immersive virtual environments. In: Proceedings of PRESENCE 2006: the 9th annual international workshop on presence, Cleveland, 24–26 Aug

Yee N, Bailenson JN (2007) The Proteus effect: the effect of transformed self-representation on behavior. Hum Commun Res 33(3):271–290

Artificial Intelligence and Machine Learning: An Instructor's Exoskeleton in the Future of Education

Stephanie E. August and Audrey Tsaima

1 Emerging Trends and Pedagogies

1.1 How Do Artificial Intelligence and Machine Learning Appear in Interactive Learning Experiences?

Technology is transforming how we solve complex problems, as well as how we share information. In this chapter, we look at an innovative learning environment from the perspectives of an enrolled student, a teaching assistant, and the professor of a fluid dynamics course with 100+ enrolled students. The scenario and research provide insight into the value of incorporating artificial intelligence and machine learning into the learning experience.

Student

Halfway through the pursuit of their undergraduate degree in chemical engineering, Alex Rhimes, age 20, from Baltimore, Mary's Lake, was planning on taking the foundational fluid dynamics class—the most notoriously difficult class in the major. The learning management system and recommendation engine used by the university suggested taking this course early based on Alex's good grades and internship experience. Like most students, Alex logged into ratemyprofessors.com before selecting the class. With ratings in

S. E. August (✉)
Department of Computer Science, Loyola Marymount University and California State University, Los Angeles, CA, USA
e-mail: august@acm.org; saugust@lmu.edu

A. Tsaima
BetterUp, San Francisco, CA, USA
e-mail: audrey.tsaima@betterup.co

© The Author(s) 2021
J. Ryoo, K. Winkelmann (eds.), *Innovative Learning Environments in STEM Higher Education*, SpringerBriefs in Statistics,
https://doi.org/10.1007/978-3-030-58948-6_5

the high 4s, Alex tabbed over to the university course site and clicked the big blue "Register" button on the screen. According to the reviews, Prof. Gomez went above and beyond to create a highly personalized environment for each student. As soon as Alex registers, an email notification is received: Pre-Course Simulation Game.

Instructor

In Charrysville, Virgonne, Dr. Riley Gomez, second year associate professor, wakes up early on the first day of class to check new emails. Rolling over in bed groggily, Prof. Gomez reaches for the phone on the bedside table, scrolling past the unfiltered junk emails. Prof. Gomez is anticipating notifications from students submitting their last-minute survey responses to the self-assessment exercise shared a week ago. As student enrollments have increased and class sizes increased, Prof. Gomez started incorporating artificial intelligence and machine learning methods into the classroom environment. It was the only feasible way to reach the 100+ students enrolled in the fluid dynamics class.

"Lecture-style classes with the sage on the stage are a thing of the past," explains Prof. Gomez. "Using algorithms is the most effective way to manage classes in the face of increased enrollment."

Prof. Gomez swipes left, left, and down in order to load the most recent results. It looks like the class is spread all over the place with experience levels and interest in fluid dynamics. It is not uncommon for students to drop out of this class and fail to persist. However, Prof. Gomez is adamant about ensuring that every student feel supported and hopeful that the simulation assessment provided an opportunity for students to learn some basics before the first session. The left chart on the dashboard shows passive traits that were monitored during the simulation. Indicators such as eye movements and facial expressions are tracked in blue and orange. The middle chart on the dashboard illustrates anticipated knowledge gaps and opportunities for support based on data mined from last year's class.

The role of artificial intelligence (AI) in US education is continuing to expand (*Artificial Intelligence Market in the US Education Sector 2018-2022—Key Vendors are Cogni, IBM, Microsoft, Nuance Communications, Pixatel, and Quantum Adaptive Learning—ResearchAndMarkets.com* 2018). As education moves toward providing customized learning paths, the use of artificial intelligence (AI) in learning systems increases, creating scaffolding that extends the ability and reach of an instructor (Tsinakos 2006) much as a physical exoskeleton combined with augmented reality enables a worker to see more than what is in front of them, and accomplish tasks they are not able to complete on their own (Srinivasan 2018). Chatbots (Bradeško and Mladenić 2012; Fonte et al. 2016; Albayrak et al. 2018; Eicher et al. 2018), autograders (Wang et al. 2018; Kyrilov 2014), and systems that passively monitor and then direct student progress (Paaßen et al. 2018) use AI, machine learning (ML), and deep learning technologies to store and process data and then communicate it to students and instructors. This exploitation of AI in education requires substantial funding and time for research, implementation, and assessment for the education community to understand the efficacy of the technology and its role in student persistence and subsequent on-the-job performance or success in graduate studies (Marr 2018; Polachowska 2019).

Artificial intelligence is the study of how to make computers perform tasks that appear to require intelligence when performed by humans. Machine learning and deep learning fall under this broad definition of artificial intelligence. **Machine learning** focuses on parsing and analyzing data in an automated fashion, without human intervention, to learn models for decision-making. Machine learning is considered a data mining technique. An algorithm that clusters data according to its similarities and differences is an example of machine learning. **Deep learning** is a subset of machine learning that relies on networks that mimic the way the human brain processes data and creates patterns to acquire decision-making ability.

A **chatbot** is a software program that converses with a human user. Chatbot ability ranges from those that conduct a shallow dialogue over a broad range of topics to those with deep knowledge and conversational ability over a well-scoped domain of discourse. The best are difficult to distinguish from human conversants. **Autograders** are software programs used to evaluate work produced by students with little or no human intervention. They can perform tasks ranging from scoring multiple choice tests to analyzing and grading essays. The findings produced by autograders range from binary (correct/incorrect) to conceptual feedback. Automatic review of essays is often combined with human review of the essays, with subsequent closer human examination of an essay if the automated and initial human results disagree. **Passive monitoring and guidance** can be integrated into an online learning system to compare a student's activities to expected behavior. An instructor might learn through the system that several students engaged in online activities appear to be making similar errors, or when a particular individual appears to be lagging behind. They can also suggest interventions to the instructor, tailored to the difficulty encountered. Likewise, these systems can provide students hints about what they might try or modules to review, as well as feedback regarding how the student is progressing relative to the rest of the class.

Applications of AI-based education technology support learning in four ways: through automated tutoring, personalizing learning, assessing student knowledge, and automating tasks normally performed by the instructor (Lu and Harris 2018).

Intelligent tutoring systems (ITS) produce statistically significant improvements in student learning outcomes, such as mastery and retention, when compared to traditional classroom teaching, independent textbook use, and non-AI computer-based instruction (Ma et al. 2014). However, experts point out that ITS curricula are rather inflexible due to technical challenges in accommodating user feedback, modified core standards, or content changes.

In addition to supporting improved student learning outcomes, the use of AI and ML in education has the potential to lead to improved teacher satisfaction (VanLehn et al. 2019a, b; Dietrich 2015). AI coupled with ML can provide 24/7 student support. It supports tracking student performance and aggregating student concerns. It can facilitate personalizing and adapting learning materials to individual students. These automated tools enable timely and passive assessment and more finely grained tracking of student knowledge and skill gains (Aleven et al. 2010; Arroyo et al. 2014). This assistance empowers the instructor, who can feel more confident in student opportunity to succeed, knowing that the students are receiving needed support that the instructor might otherwise struggle to provide. The instructor is able to devote time to creative activities and feedback beyond what the automated systems can provide, such as affective feedback and support (Wu et al. 2016; Duo and Song 2012).

The implementation of an online learning system requires a sophisticated digital ecosystem that incorporates the complex interactions among students, instructors, and content. It must include a sophisticated human-computer interface that supports access, monitoring, feedback, and assessment (Reyna 2011; Rezaei and Montazer 2016). The system can be built upon an existing e-business solution or learning management system, or arise from an array of independent modules. These systems are often cloud-based, providing services over the Internet, to provide maximum accessibility.

2 Use of AI and ML in 2026

2.1 Content Presentation

2.1.1 Opportunities

1. **Multifaceted presentation**, such as mobile computing, the Internet, natural language interfaces, gesture-based interfaces (Audinot et al. 2018; Bowman et al. 2008; Case 2018), silent speech recognition (Waltz 2019), and other advances in tools for user interface development have led to a richness in the modes of communication between humans and machines. Guided learning (Chi and Barnes 2014; Price et al. 2016; Ontañón et al. 2017) automates feedback tailored to student needs and gives students control over their learning (Zhou et al. 2016; Harackiewicz et al. 1987). Brownfield programming (Baley and Belcham 2010; Vujičić et al. 2018) allows students to learn programming by studying legacy

systems. Simulation-based learning (Lateef 2010; *Official Site | Second Life— Virtual Worlds, Virtual Reality, VR, Avatars, Free 3D Chat* 2019; *OpenSimulator* 2019; OPNET Optimum Network Performance 2020) offers alternatives to real-world experiences. Content can be presented through multiple modalities, such as text, video, audio, or simulation. These advancements are extending our ability to interface with machines and each other. However, more work is needed for such systems to become highly reliable, robust, and widely available (Case 2018).

2. **Simulations** that embed AI and ML through interactive, realistic recreations of real-world scenarios provide additional means for presenting content. These allow students to gain experience with environments that would otherwise be inaccessible to them. Examples include high-performance computing and networking, in which access to commercially available or real-world systems would be costly or pose privacy or security concerns. Simulators, such as the OPNET network simulator (OPNET Optimum Network Performance 2020), simulate the behavior and performance of any type of network. To have a broad impact, a simulator must be easily ported to new environments or shared with low cost.

3. Content is also provided when students receive **automated feedback** on their work in progress. This includes automated grading of and feedback on assignments as student complete exercises (Kyrilov 2014; Wang et al. 2018), chatbots that respond to student questions (Eicher et al. 2018), or automated delivery of hints that guide a student's next step in solving a problem (Paaßen et al. 2018). Grading rubrics based on a case-based reasoning paradigm reusing feedback on previous similarly graded coursework is an additional means of supplying content (Wiratunga et al. 2011).

2.1.2   Challenges

1. **Content management** is a challenging problem yet to be mastered. The more data that is accumulated, the greater the level of curation that is required. Intelligent storage and retrieval are needed to enable the presentation of context-relevant content (Miller 2017). This problem is being tackled by means such as observing the performance of a student over time and adjusting the type and frequency of automated hints provided to students by a behind-the-scenes passive assistant (Mostafavi and Barnes 2017; Peddycord-Liu et al. 2016). Each modality of presentation requires a different level of expertise for creation, delivery, and curation. Accessibility is also a challenge, raising concerns regarding how to present the same content in different modalities while accounting for various student (dis)abilities. Assignments and other activities clearly linked to learning objectives and tied to curricular requirements need to be both available and curated (Akbar 2013). Professional development on simulators and other tools and supporting documentation need to be available, as well.

2. **Feedback** provided via autograders ranges from binary responses indicating correct or incorrect to conceptual feedback. Detailed feedback enhances the

learning experience (Kyrilov and Noelle 2016). There is evidence that instant binary feedback increases the likelihood that students will cheat on assignments (Kyrilov and Noelle 2015). Conceptual feedback can be provided through case-based reasoning. A case-based reasoner can analyze student errors, compare them to previously recognized errors, and retrieve and tailor specific guidance that proved useful to other students (Wiratunga et al. 2011; Kyrilov 2014).

2.1.3 Implementation Strategies

Content delivery can be cloud-based, server-based, or a combination of both. Fuad, Akbar, and Zubov present Dysgu (Fuad et al. 2018a, b) as an example of a cloud-based interactive learning environment. Dysgu personalizes and adapts out-of-class activities to satisfy individual student needs. Dysgu employs mobile technology to present activities that are smaller than traditional out-of-class activities. It incorporates social networking, which supports anonymous interaction and allows students to gauge their progress relative to the progress of other students.

In contrast, Isomöttönen, Lakanen, and Lappalainen's TIM (The Interactive Material) is a document-focused system (Isomöttönen et al. 2019). TIM's document-oriented user interface supports creating and editing learning materials, rather than managing course content. Instructors are able to track document sections (un)read by individual students. It also supports automated assessment, gamified learning through performance monitoring and display, and comprehensive tracking of submissions and user interactions with the system.

2.1.4 Research Questions

The appropriate use of online learning systems, their curation, and assessment are open areas. Several concerns need to be addressed on the path to full and efficient integration with learning experiences:

1. **Maintenance**. How will content be curated over time?
2. **Ownership**. Who will curate content over time?
3. **Relevance**. What is an effective mapping of tools to learning situations?
4. **Effectiveness**. Which data most accurately reflects the effectiveness of intelligent content delivery beyond measuring content knowledge?
5. **Affect**. Are affective measures more important? Do they reflect persistence, retention, or later mastery?
6. **Customization**. What functionality is needed to make assignment creation easier, less time-consuming, and flexible, allowing instructors to customize material to fit student level, knowledge, culture, and institutional requirements?
7. **Detail**. What level of detail and what type of information do students need to maximize the learning experience?

2.2 Interactions and Communications

Classroom

Five minutes before the first class, Dr. Gomez logs into room system to turn on the affective computing machine. The cameras in the classroom quietly swivel toward the students' seats and begin populating data to the instructor screen regarding individuals' moods. Taylor Speek, the 27-year-old teaching assistant, starts meticulously drafting learning plans for students that are showing signs of poor engagement and difficulty grasping knowledge during the first session, making decisions based on an analysis of previous courses related to fluid dynamics. This class is particularly rich in information because all the students have signed waiver forms to be recorded and have their physiological and physical traits captured to monitor and notice patterns and trends in their progress. Looking at the pre-course assessment, Taylor notices that 83% of the learners showed signs of stress according to the data on their wearables, particularly in the area on Bernoulli's principle. Taylor adds demystifying Bernoulli's principle to the class agenda and gives Prof. Gomez a heads up.

During the class, Alex runs into more confusion about Bernoulli's principle and makes mouth movements so that her silent speech recognition device picks up the question without disrupting the class. "It makes you feel less self-conscious," argues Alex. "My older sibling went into college thinking they would also pursue chemical engineering but then kept failing classes because they couldn't get help and felt too embarrassed to ask." The question is funneled to the class chatbot, which provides helpful resources curated by Prof. Gomez. The affective computing software on Alex's computer registers that although the resources alleviate some of the stress, this student will need more support after class. The same question which has been asked in similar ways throughout the class is anonymously posted as a single topic to the class community for discussion afterward. Prof. Gomez proceeds with the final activity seamlessly, knowing that the peer learning forum used will foster more opportunities for students to reinforce what they learned from each other today. The chatbot reports back to the mainframe and automatically schedules a session between Taylor and Alex when they are both available.

2.2.1 Opportunities

1. **Interactions and communications** in an intelligent learning environment span user-initiated information retrieval and question answering, facilitation of informal or directed peer-to-peer communication, and student-facilitator interactions for guided learning. The facilitation of natural language interactions with systems, from storytelling to question answering, has been studied since

the birth of artificial intelligence in the middle of the twentieth century (Winston 2016). Theories were devised for inferring the appropriate response to a question (Lehnert 1977; Wilensky 1977), and systems were constructed to allow novice users to ask about complex systems, such as Wilensky's UNIX Consultant (Wilensky et al. 2000). Later systems supported natural language interfaces to databases (Androutsopoulos et al. 1995) and ultimately IBM's Watson suite, designed for building conversational interfaces into any application, device, or channel (IBM 2018). Despite these advances, these interfaces lack deep knowledge and instead rely on massive amounts of data to train the system. While this is useful in most situations, being able to handle novel queries or mundane queries posed in an unusual fashion remains a challenge.

2. **Informal communication and collaborative problem-solving** are essential components of any learning experience. Students need to be able to develop a sense of community with their peers, as well as have access to expert knowledge and guidance from the instructor and teaching assistants. Dashboards allow the instructor to monitor individual student activity, present aggregate feedback on student activity, inform the instructor when a student appears to be having difficulty, and offer suggestions on additional approaches to presenting material that has the potential to enrich the learning environment.

3. Informal learning and co-curricular activities can be enhanced with **guided linear learning** tailored to a specific student's needs and learning objectives. Fuad et al. (2018b) and Akbar (2013) use gamification to this end in their project "Active Learning for Out-of-class Activities to Improve Student Success." In addition, points of intervention can be automatically identified, and the system can respond in collaboration with the instructor. Chen et al. (2019) develop this instructional strategy in their system that automatically delivers prompts to students based on the comments submitted when they commit software code to a source code repository. Their reflection-in-place app shown in Fig. 1 builds on a recommender system and guides students to reflect on their work in a meaningful way.

2.2.2 Challenges

Affect is an important component in human-human communication and can significantly influence learning experiences (Wu et al. 2016). Affective computing has three components: detecting the emotions of the user, expressing what a human would perceive as an emotion, and actually experiencing an emotion (Picard 2003).

(1) and (2) Accurately **detecting** and **conveying** appropriate emotions is a complex task that is not well addressed in current intelligent systems.

(3) The **integration** of emotions, mood, motivations, and personality contributes to user engagement (Fatahi and Moradian 2018). The lack of these dimensions is

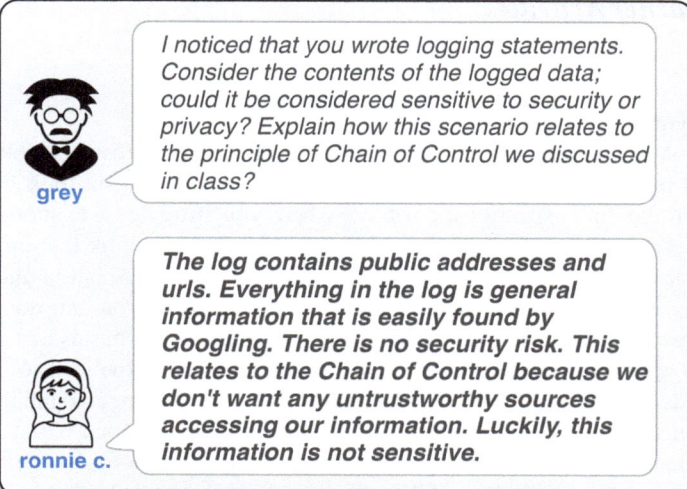

Fig. 1 An example of Chen, Ciborowska, and Damevski's automated reflection-in-action system. A student (ronnie c.) developing a secure mobile app is encouraged by the recommender system (grey) to learn about the security principle of chain of control (Reproduced from Chen et al. 2019)

thought to lessen learner satisfaction with e-learning systems and lead to a higher drop rate of online courses. The challenges lie in the sophistication of the systems, the complexities of intercultural communication, and developers' awareness of affect and ethics (Cowie 2014).

2.2.3   Research Questions

1. **Adaptation**. What is the appropriate mechanism for adaptation?
2. **Assistance**. What is the appropriate level of assistance to give students at various stages of learning?
3. **Dynamic adjustment**. Does the instructor have the ability to dynamically increase or decrease assistance?
4. **Automatic adjustment**. Can the level of assistance be automatically adjusted based upon a student's performance?
5. **Affect**. What is the impact of multimodal emotion recognition and corresponding emotional response in e-learning systems?
6. **Ethics**. What is the role of ethics in human-computer communications and affective computing? What is an effective strategy for preparing the next generation of developers and educators to incorporate access and ethics into product design and content delivery?

2.3   Learner Activities

Intervention

During Alex's support session, Taylor comments on an interesting tidbit high-lighted by the data dashboard. "Alex, it looks like when you took Process Control, the only aspect of the subject where you struggled was momentum balances, which relates to Bernoulli's principle. Do you think that might be why you're struggling?" Alex's face lights up with this revelation and head nodding ensues. "Let me assign this additional activity you can do during your free time. I think it'll help a lot with coming back to basics and slowly ramping back up. I think you should try and do this before the next session. It scaffolds the level of difficulty as you go and should challenge you in the cor-rect areas. This app will customize the content to be more video-based since I know that's what you seem to prefer. Is that correct? And don't worry, a lot of students struggle with Bernoulli's principle." Taylor spends the remaining 15 minutes they have together providing emotional support, answering ques-tions, adding points of clarification, and ensuring that Alex receives the addi-tional materials. As the student leaves the room, another one takes his place, and Taylor's dashboard changes to the appropriate student. When meeting with her master instructor, Taylor reflects, "These tools have made my work so much more meaningful and efficient. I get to support more students than I did in the past and really target the key areas of need."

In the past decades, student activities and course assignments have evolved in new directions, from traditional take-home work to flipped classrooms (*Flipping the Classroom* 2019), from one-size-fits-all curriculum to personalized learning, and from paper and pen to digitized submissions requiring broadband Internet and access to personal devices. With these advancements, institutions have adapted their curriculum to more finely deliver and assess learning experiences more focused on personalization than in the past—creating an environment that is ripe for leveraging machine learning models. As such, the lines between activities and assessments have also blurred over time. In some cases, instructors become facilitators of experi-ences as opposed to disseminators of information. In this section, we explore the current opportunities, challenges, implementation strategies, and further research questions through the lens of uncovering best practices in intelligent instruc-tional design.

2.3.1 Opportunities

1. **Interactive** and adaptive game-based learning is an increasingly popular area of investment for researchers focused on designing new digitized and innovative learning activities, which are often examples of constructivism in education. Constructivism explains that humans acquire and gain knowledge through experiences (Wadsworth 1996). LewiSpace by Ghali et al. (2016) is an example of an exploratory educational game that was developed with Unity 4.5 to teach a college-level chemistry lesson on drawing Lewis diagrams, which are structural representations of molecules. LewiSpace captures learners' physiological traits such as electroencephalography (the measurement of electrical activity in different parts of the brain and the recording of such activity as a visual trace), facial expression, and eye movements throughout the game. LewiSpace also pulls data from a personality traits questionnaire in order to determine a learner's performance and potential need for help. Experimenting with multiple machine learning algorithms, Ghali et al. (2016) found the highest level of accuracy in predicting failure rates using a logistic regression model.[1] Future versions of LewiSpace will incorporate real-time measurements to adapt the experience to learner's needs.

   Similar to Dysgu (Fuad et al. 2018b), an interactive mobile game that is currently manually scaffolded by instructors, and Epplets (Kumar 2018), an interactive tool for solving Parsons puzzles (Kumar 2017), LewiSpace presents a typical case study of the potential for simulations to be designed in more adaptive manners for the enhancement of constructivist learning. To support such efforts, Li et al. (2010) developed an adaptive course generation framework, which extracts course materials and learner profiles to help instructors design courses that are more advanced in handling multiple learning characteristics such as style and preference. This tool helps facilitate the alignment of course curriculum with the creation of adaptive activities.

2. **Supportive** elements in learning activities are crucial to ensure a relatable experience. Particularly in problem-based learning, enhancing critical thinking skills outside of the classroom may take the form of discussion-based activities. In such instances, intelligent learning tools such as MALESAbrain (Chiang and Fung 2004) encourage learners to judge their peers' solutions before exploring further content. This information is used to rank and arrange learning issues in an effort to transform obligatory forums and chat rooms into rich discussion opportunities. This provides an opportunity to improve the quality of conversations inside and outside of the classroom, which may increase the opportunities for peer-to-peer learning.

---

[1] In statistics, the logistic model is used to model the probability of a certain class or event existing such as pass/fail, win/lose, alive/dead, or healthy/sick.

2.3.2 Challenges

There are many remaining challenges toward leveraging artificial intelligence in regard to learner activities. Important challenges include:

1. Physiological traits such as facial expressions may be promising avenues for understanding and measuring engagement, but **interactive** and adaptive simulations experience obstacles when using metrics such as pupil dilation and emotions for predicting learner success (Ghali et al. 2016) due to the stimulating nature of the experience. These indicators may often add more noise in the form of irrelevant information or randomness in a dataset used by machine learning models trying to extract and determine specific causality, though future developments may overcome this sensory obstacle.

2. **Access** to personal devices that can successfully accommodate technological needs remains one of the most challenging obstacles of deploying learner activities. The digital divide persists for students within certain educational institutions and communities (Digital Divide Compounds U.S. Education Equity Problem, First-of-Its-Kind Survey Reveals 2018). As learner activities become increasingly demanding on devices and connectivity, ensuring that all students are equipped with the necessary means to access content will require careful attention. In order for engagement in learning to exit the classroom, simply equipping institutions will not suffice.

2.3.3 Implementation Strategies

Throughout the US education system, learner activities are often disseminated as instructors see fit to fulfill curriculum requirements and learning objectives. As such, instructors are a crucial part of the process for successfully deploying, tracking, and triaging these activities. As learning activities become more adaptive and personalized than instructors themselves, human instructors often become facilitators of experiences as opposed to disseminators of information.

Epplets (Kumar 2018) are a examples of software assistants designed to help students working alone learn good programming principles and algorithm design. The instructor retains the role of passing along knowledge to learners before the activity has begun. This small-scale integration of intelligence is an approachable first step in observing the outcome of allowing students to practice and assess at their own pace and level of rigor. Epplets enables a teacher to maintain a relatively hands-on approach in monitoring progress and providing aid where necessary. In more immersive experiences such as LewiSpace (Ghali et al. 2016), which replace an entire lesson, including the process of disseminating knowledge, instructors should be prepared to smooth the transition between classroom and simulation. In-person class time should be focused on deconstructing the virtual interaction, facilitating discussion, and processing learnings into applicable knowledge outside of the experience.

Instructors within institutions should provide a fluid and seamless experience inside and outside of the classroom for learners, particularly for flipped experiences.

2.3.4   Research Questions

There are a number of opportunities for further study. A few key examples include:

1. **Immersion**. How might we make learning experiences more immersive while maintaining transferrable real-world skills?
2. **Decision-Making**. How might instructors remain key decision-makers in adaptive learning experiences that often adjust with only learner data inputs?
3. **Indicators**. What might be the best measurements of competency and engagement for simulation-based activities?
4. **Noise**. Which psychological traits and data sources are the best indicators of learning and engagement?

2.4   Assessment

Assessment
At the end of the semester, Prof. Gomez wraps up the last class with an essay-based exam. Students are asked to answer questions based on their greatest opportunities for improvement. Alex receives three prompts on inviscid flow and one on Bernoulli's principle. The machine learning model is fine-tuned to Prof. Gomez's competency-focused goal of measuring improvement to ensure that the students are well-rounded and confident in their abilities. "Why test a student on a topic that I know they're an A+ on?" shares Prof. Gomez, "Let's get straight to the point, what I care about is whether all of their skills are up to par and that we're being effective in our delivery of learning. Throughout the course, we should have caught all of the pain points and now we're just confirming." As soon as Alex clicks "Submit," the essays are automatically graded with a natural language processing tool. Prof. Gomez and Taylor also receive a new report on their dashboard, indicating that this final piece of data from the course has increased Alex's likelihood of successfully graduating in chemical engineering.

A crucial component of determining the effectiveness of any learning program, intervention, and/or activity is being able to meet or exceed anticipated student learning outcomes. In optimal situations, all students would be equipped to successfully accomplish a variety of learning goals within and outside of traditional learning environments. However, it is common knowledge among students, educators, and parents that this is simply not the case. Leveraging artificial intelligence and, more specifically, machine learning methods can illuminate opportunities to quickly assess student performance, provide accurate feedback, proactively engage with students to an extent not possible without the intelligent intervention, and predict likely student outcomes. In the succeeding paragraphs, we explore the existing literature and opportunities within four major categories: (1) predicting performance, (2) reading and writing tasks, (3) zone of proximal development, and (4) personalized learning.

2.4.1 Opportunities

1. **Predicting Performance** is an essential area of study for machine learning applications in education. The traditional approach to monitoring student performance is to make assessment scores central to determining student achievement (National Research Council, Division of Behavioral and Social Sciences and Education, Center for Education, Board on Testing and Assessment, and Committee on the Foundations of Assessment 2001). To date, using these assessments to determine learner outcomes has been challenging due to ineffective and inefficient testing. Ogor (2007) proposes a methodology with a 94% success rating for monitoring students' performance and predicting graduation status by capturing continuous assessment and examination scores. Alternatively, Ciolacu et al. (2017) have an interesting and novel approach to estimate student performance at examination through analyses based on neural networks, support vector machines, decision trees, and clustering. In this work, the authors leverage a blended learning course and a complete virtual course to test their model's prediction accuracy. Another performance-related application of machine learning is predicting teamwork effectiveness by extracting objective and quantitative team activity data (Petkovic et al. 2012).

2. **Reading and Writing Tasks** are ubiquitous activities that all students encounter in higher education. EdX (*EdX: About Us* 2013), a massive online open course (MOOC) provider, has created a machine-based automated essay scoring (AES) application to assess student work at scale (Balfour 2013). Martinez et al. (2013) describe an AES that utilizes support vector machines, software encompassing supervised learning models for data classification and regression analysis. An AES must calibrate for each writing assignment and grade-level. They have been shown to correlate more highly with human raters than human raters among themselves (Shermis et al. 2010). AES offers immediate feedback to students though it is limited by unique speech elements such as humor. Nehm et al. (2012) developed another method with a high success rate. They created a "summarization integrated development environment" program, which assesses written explanations in biology using natural language processing. Assessment of a student's reading level in order to improve the teacher's ability to support an individual student's learning is another opportunity to leverage support vector machines (Petersen and Ostendorf 2009). Although this application may be more

useful in primary education, these principles may still support higher education English-as-a-Second-Language (ESL) students in all fields.

3. Vygotsky's **Zone of Proximal Development (ZPD)** is an important concept that refers to the difference between what a learner can accomplish with help and by themselves (Chaiklin 2003). Traditional models of tutoring, office hours, etc. are difficult to scale and generally triggered by low grades, which occur after a student has an unsuccessful learning experience. Ahadi et al. (2015) explore machine learning methods, which automatically identify students in need of assistance by observing constantly accumulated early data such as students' progress on assignments and behaviors in lectures. In an interesting contrast, Beck et al. (2008) investigate through the Bayesian evaluation and assessment framework whether or not tutorial interventions actually help students improve their outcomes and develop long-term and translatable skills.

4. **Personalized Learning** is an increasingly opportunistic challenge as learning at scale becomes more prevalent. How might learners improve their outcomes when in-person and virtual class sizes grow? García et al. (2007) focused on detecting students' learning styles by evaluating the precision of Bayesian networks. The authors' model infers a student's style by capturing aspects of human behavior while the student is working with the system. Instead of assessing human behavior, Blikstein (2011) was interested in predicting it in the context of open-ended environments when performing tasks such as computer programming. AdaLearn (Alian and Al-Akhras 2010) creates a profile of learner responses to use for recommending content to learners. RubricAce (Wiratunga et al. 2011) improves rubric-based feedback to students using a case-based paradigm.

Finally, affective computing provides a promising avenue for machine learning applications as learner engagement, knowledge retention, and many other components of learning are influenced by emotion. Wu et al. (2016) provide a review of current trends and challenges with affective computing in education and learning. The authors identify common data collection and machine learning methods used. They also highlight opportunities to leverage insights to intervene at appropriate moments to improve learner trajectories.

2.4.2 Challenges

There remain a number of outstanding challenges toward fully developing artificial intelligence within synchronous and asynchronous environments. Key challenges in the four sections outlined above include:

1. **Predicting performance** (Ogor 2007; Ciolacu et al. 2017) presents ethical challenges if learners are given opportunities and support based on their anticipated grades and test scores from machine learning algorithms. Measures of performance on tests and grades are limited definitions of student achievement and real-world outcomes in STEM fields (Spector 2017). This approach limits the

way in which we define learner success and evaluate whole-person skills that produce effective STEM graduates such as resilience, grit, or lifelong learning (Strauss 2017).

2. **Reading and writing tasks** that leverage AI require machine learning models to be trained for each assessment deployed at each grade level (Balfour 2013; Shermis et al. 2010; Nehm et al. 2012). This can be time-consuming and potentially costly for institutions that may not have the technical resources to implement AES tools. For others who may seek to purchase out-of-the-box solutions, this can create a black box where the AI is not explainable (Knight 2017). Instructors and administrators will be unable to clearly understand how written assignments are being graded.

3. Identifying appropriate interventions to help students reach their **zone of proximal development** requires a multitude of inputs, which may often be missed even by instructors in traditional classrooms. Predictive models created for specific situations may not be applicable and transferrable to other contexts, which may have variability in many aspects, such as teaching approach, materials, or group of students (García et al. 2007). An important part of this challenge arises from the difficulty of recognizing and accounting for these alterations in order to adjust the models being used.

4. **Personalized learning** powered by affective computing provides an opportunity to tap into human characteristics through facial recognition and other physiological traits (Wu et al. 2016). Unfortunately, there are a limited number of emotions that can be accurately identified, and there are concerns regarding the applicability of these defined traits to all demographics (Do 2019).

Several important questions remain, especially in the area of societal considerations. Can an AI be programmed to accurately account for racial, cultural, and religious differences? Can this be accomplished without controversy or running the risk of inappropriate racial or other profiling? Are there basic, humanistic beliefs that can be integrated into these systems to ensure equitable treatment and respect for all? Finally, there is the issue of consent. What challenges does a mixed classroom of those who consented and those who did not create? Does such a mixed classroom affect outcomes on both sides?

2.4.3   Implementation Strategies

In many cases, AI for assessment and evaluation are embedded within standard synchronous and asynchronous activities, e.g., evaluating test scores or written work. Data mining techniques are leveraged with machine learning models to passively determine desired predictions. As universities invite more remote students into their programs, there is an increased amount of potential data to be gathered, tested, applied, and analyzed. In instances such as Dysgu (Fuad et al. 2018b), instructors have manually assigned learner activities (replacing the judgment piece of what may become AI in the future) in order to increase student engagement outside of the classroom. This experiment provides a low-fidelity first step for institutions to test whether or not implementing AI in certain areas would be impactful to a student's learning outcomes.

Additionally, the partnership between AI and the instructor is a key relationship to balance. Fully applying AI in a classroom or within institutions will require clarity in terms of role definitions and leveraging each party's strengths. For instance, an instructor may be more effective in providing an emotion-based intervention after a machine learning algorithm has identified a disengaged learner, rather than having the instructor initiating an emotion-based response without understanding the current affective state of the learner.

2.4.4 Research Questions

There are a multitude of opportunities for future study. A few crucial examples include:

1. **Bias**. How might we train machine learning models to avoid replicating in-classroom, instructor, and institutional biases toward assessing certain demographics?
2. **Accessibility**. How might we leverage AI assessments in-classrooms and outside of classrooms to enhance all students' sense of self-efficacy in STEM fields?
3. **Interactions**. Where might AI evaluations be best served to measure and improve student outcomes?
4. **Workplace Skills**. How might we leverage AI to better train, assess, and prepare STEM learners to thrive in the global workforce?

2.5 Co-curricular Activities

Learning is being transformed by intelligent systems (Schmelzer 2019). Co-curricular activities are those which relate to and support an academic course of studies. AI-driven co-curricular activities are an opportunity to support online learning, the various ways people learn, and the rates at which they learn. Machine learning-based development of student profiles and customization of training materials allow instructors to draw upon a single curriculum while modifying content and presentation for individual users. Online textbooks and their interactive interfaces further support tailoring of content and personalization of delivery and feedback. They facilitate providing students hints as they work through assignments and conceptual feedback and as exercises are automatically assessed.

A learner interacts with material in many ways external to the formal educational event. Thus, the student experience consists of a wide range of interactions that may consist of athletic, scholarship, social, and service dimensions. Providing students with access to a range of these dimensions is a critical link in developing the overall quality of the student experience. Workshop participants identified several forms of interaction. They called for the augmentation of tutoring with personal, conversational education assistants, often referred to as autonomous conversational agents (IBM 2018). These and other integrated intelligent autonomous education agents must respond effectively to a learner's questions and provide assistance with learning or assignment tasks. Learning systems need to reinforce concepts in a personalized fashion with additional materials to reinforce the curriculum. Furthermore, they need to allow students to learn at their own pace, to satisfy their own goals. Co-curricular education involves many opportunities and challenges. Co-curricular activities are implemented in a variety of ways and result in several open research questions.

2.5.1 Opportunities

X-FILEs workshop participants identified several characteristics of future learning and future students. Their consensus was that the basic pedagogy will still be delivered with new technology, and predicted that the best pedagogical ideas will be more fully realized. The Internet facilitates delivery of content to large numbers of individuals both synchronously and asynchronously, and the scale of delivery is expected to change. Feedback will become more automated, and content and delivery will be tailored to instructional objectives and the individual learner's interests and needs. As institutions are called to do more with limited budgets, the importance of virtual environments over brick-and-mortar settings will increase.

Participants were asked how the teaching and learning process might be different in the future. Four responses stood out:

- Students will need to be more responsible for keeping up with the class. As a result, self-motivation will become essential.
- Access to educationally valuable locations will be freed from temporal and spatial constraints.
- Students will become more accountable for conducive learning.
- Learning will become more active, less passive.

A classroom environment that is conducive to learning entails staging the physical space, creating a communal environment, maintaining a positive climate and culture, and, most critically, convincing the students to become cooperative, active learners (Lynch 2016).

2.5.2 Challenges

Several challenges must be addressed to implement widespread adoption of automated intelligent co-curricular activities. These can be categorized as seeing a need for the systems, as well as acceptance, availability, assessment, and robustness.

1. **Need** Smaller programs that pride themselves on small classes and instructor-led courses might not recognize the value of automated intelligent co-curricular activities, or be reluctant to adopt them for fear of tarnishing their image of instructor-led, student-focused environments. In the age of COVID-19-motivated online learning, instructors are often encountering student questions that normally would be answered by readily accessible teaching assistants. These instructors are beginning to understand the need for automated activities.
2. **Acceptance and Availability** Social acceptance of automated agents that support co-curricular activities can present a challenge among both faculty and students. Workshop participants suggested that awareness of the value to student learning and independence and the availability of co-curricular activities can be raised through campus wide initiatives, such as panels, research symposia, industry, and examples. To increase adoption, one workshop participant advocated for holding a competition for faculty and students for the purpose of identifying instances of AI in the local institution's environment. A leaderboard was proposed for recording, for example, the most creative entry or promising ways to increase student motivation, with prizes offered for the best examples. Identifying individuals on campus with experience and expertise and experience with automated co-curricular systems is essential when they are first introduced. Staff in an Instructional Technology office can prove invaluable to the rollout of this technology.
3. **Assessment** The efficacy of the co-curricular activities needs to be assessed. To accomplish this, institutional data from the library or institutional research unit could be made available to the community to build algorithms and related applications to measure the effectiveness of the activities.
4. **Robustness** Our students represent many abilities, countries, nationalities, communities, and cultures in many time zones, on many schedules, and with varying degrees of Internet connectivity. To be effective and accepted by a wide audience, the challenges of this diverse audience and these diverse environments need to be addressed. Systems need to be compliant with accessibility standards (US EPA 2013). Developers need to be aware of unconscious biases unintentionally embedded in systems that can misunderstand or alienate users (Eicher et al. 2018). In addition, a robust system would be able to understand and respond and present information in a variety of natural languages and at varying levels of abstraction (August 2012).

2.5.3 Implementation Strategies

AI-based co-curricular activities range from games to tutors to immersive simulations. Gamification platforms such as Classcraft, Rezzly, Seppo, Youtopia, and Kahoot! offer external motivation, such as rewards and leaderboards, and internal motivation, such as autonomy and mastery, to engage students in meaningful learning experiences (Goshevski et al. 2017). The instructor is able to provide or tailor content to increase relevance to the target content. AI chatbots, such as Jill Watson at Georgia Tech (Eicher et al. 2018), as well as Mtabe from Tanzania and LangBot in Ethiopia (Nsehe 2019), provide students personal tutors, tailored to instructional and learner needs.

Adaptive learning has the potential to promote access and quality at scale in higher education (Becker et al. 2018). Cavanagh et al. (2020) lay out the design of framework for adaptive learning and best practices for its use. The features of the design framework include objective-based learning knowledge units, personalized assessment and content, adaptive learning paths, alternate content, and procedurally generated questions. These systems can be implemented as standalone components or integrated into an existing learning management system. ALEKS (*Overview of ALEKS* 2020; Boyce and O'Halloran 2020) is one widely used for algebra. M-Shule from Kenya (Haba 2017) is an example of a data-driven personalized learning system for K-12 education.

Virtual labs present another opportunity for co-curricular activities that involve immersive simulations. Labster (*The Complete Guide to Virtual Labs* 2020) and PraxiLab (*Virtual Science Labs at Your Fingertips* 2020) offer commercially available virtual labs for secondary and higher education, as well as other informal learning needs.

Immersive platforms such as SimCity®, Second Life®, and the *Unity* real-time development platform are additional opportunities for implementing co-curricular activities that allow 24/7 access to engaging activities that support informal learning (August et al. 2016); Winkelmann et al. 2017).

2.5.4   Research Questions

Workshop participants offered a number of open research questions related to intelligent autonomous education agents in co-curricular activities:

1. **Implications of AI**. What are the implications of AI across co-curricular areas for the implementation and use of ML?
2. **AI enhancements**. How can AI expand the efforts in co-curricular activities?
3. **Levels of formality**. How do formal concepts and contexts differ from informal concepts and contexts?
4. **Human vs. machine intelligence**. What does comparing human learning and human intelligence to machine learning and machine intelligence tell us about what it means to be human?

3   Conclusions

Development of robust, engaging, effective digital systems for learning must engage all classes of stakeholders from conception through implementation and evaluation. Such systems need to be integrated into the learning environment and into the routines of the instructors and students. Discussants at the 2018 X-FILEs Workshop identified several concerns:

- What is the relationship between the use of AI and ML in augmented learning systems and concerns such as ethics, empathy, equity, collaboration, and positive social change? Is there an obligation to consider them in parallel with the development of intelligent systems?
- What are the metrics and observations that would provide the greatest insight into the impact of AI in learning systems? Are the required data immediately available, or do they require longitudinal studies?

Addressing the first concern requires broad studies of innovative learning environments over diverse demographic groups and a range of higher education institution types. These will become more feasible over time as interactive learning environments are more widely adopted. Addressing the second concern requires looking to the longer term beyond gains in content knowledge and examining the affective impact of these learning opportunities, as well as development of critical thinking skills and fostering independent learning.

Many other questions remain to be considered:

- What are best practices for rolling out comprehensive online learning systems to ensure successful integration and achievement of learning objectives?
- What is the role of an intelligent online learning system in primary school? Secondary school? Higher education?
- How is an intelligent online learning system best integrated into primary education? Secondary education? Higher education?
- What is the role of the instructor in each?
- How are student/teacher interactions best integrated?
- What concerns do/should people have regarding limits on screen time, especially for younger students (Marr 2018)?

A rollout of the Summit Learning Platform (2017), a Chan Zuckerberg Initiative, points to multiple areas for future study, including integration of student/instructor integration, appropriate limits on screen time, parent acceptance, access to vetted resources, and controlled access to non-vetted resources. Community experiences in a Kansas school district reflect the need for more thought on these points before successful integration and achievement of learning objectives can be achieved (Bowles 2019).

References

Ahadi A, Lister R, Haapala H, Vihavainen A (2015) Exploring machine learning methods to automatically identify students in need of assistance. In: Proceedings of the eleventh annual international conference on international computing education research—ICER '15, pp 121–130. https://doi.org/10.1145/2787622.2787717

Akbar M (2013) Integrating community with collections in educational digital libraries. PhD Thesis, Virginia Polytechnic Institute & State University

Albayrak N, Ozdemir A, Zeydan E (2018) An overview of artificial intelligence based chatbots and an example chatbot application. In: 2018 26th signal processing and communications applications conference (SIU), pp 1–4. https://doi.org/10.1109/SIU.2018.8404430

Aleven V, Roll I, McLaren BM, Koedinger KR (2010) Automated, unobtrusive, action-by-action assessment of self-regulation during learning with an intelligent tutoring system. Educ Psychol 45(4):224–233

Alian M, Al-Akhras M (2010) AdaLearn: an adaptive e-learning environment. In: Proceedings of the 1st international conference on intelligent semantic web-services and applications—ISWSA '10, pp 1–7. https://doi.org/10.1145/1874590.1874611

Androutsopoulos I, Ritchie GD, Thanisch P (1995) Natural language interfaces to databases—an introduction. Nat Lang Eng 1(1):29–81. https://doi.org/10.1017/S135132490000005X

Arroyo I, Woolf BP, Burelson W, Muldner K, Rai D, Tai M (2014) A multimedia adaptive tutoring system for mathematics that addresses cognition, metacognition and affect. Int J Artif Intell Educ 24(4):387–426. https://doi.org/10.1007/s40593-014-0023-y

Artificial Intelligence Market in the US Education Sector 2018-2022—Key Vendors are Cogni, IBM, Microsoft, Nuance Communications, Pixatel & Quantum Adaptive Learning—ResearchAndMarkets.com (2018, August 27). https://www.business-wire.com/news/home/20180827005505/en/Artificial-Intelligence-Market-Education-Sector-2018-2022%2D%2D

Audinot A, Goga E, Goupil V, Jorqensen C-J, Reuzeau A, Argelaguet F (2018) Climb, Fly, stack: design of tangible and gesture-based interfaces for natural and efficient interaction. In: 2018 IEEE conference on virtual reality and 3D user interfaces (VR), pp 856–857. https://doi.org/10.1109/VR.2018.8446244

August SE (2012) Enhancing expertise, sociability, and literacy through teaching artificial intelligence as a lab science. pp 25.569.1–25.569.11. https://peer.asee.org/enhancing-expertise-sociability-and-literacy-through-teaching-artificial-intelligence-as-a-lab-science

August SE, Hammers ML, Murphy DB, Neyer A, Gueye P, Thames RQ (2016) Virtual engineering sciences learning lab: giving STEM education a second life. IEEE Trans Learn Technol 9(1):18–30. https://doi.org/10.1109/TLT.2015.2419253

Baley K, Belcham D (2010) Brownfield application development in .NET [Book]. Manning Publications. https://www.oreilly.com/library/view/brownfield-application-development/9781933988719/

Balfour SP (2013) Assessing writing in MOOCs: automated essay scoring and calibrated peer review™. Res Pract Assess 8:40–48

Beck JE, Chang K, Mostow J, Corbett A (2008) Does help help? Introducing the Bayesian evaluation and assessment methodology. In: Woolf BP, Aïmeur E, Nkambou R, Lajoie S (eds) Intelligent tutoring systems. Springer, Berlin, Heidelberg, pp 383–394

Becker SA, Brown M, Dahlstrom E, Davis A, DePaul K, Diaz V, Pomerantz J, EDUCAUSE, New Media Consortium (2018) NMC Horizon Report: 2018 Higher Education Edition. EDUCAUSE. 4772 Walnut Street Suite 206, Boulder, CO 80301-2538. Tel: 303-449-4430; Fax: 303-440-0461; e-mail: info@educause.edu; Web site: http://www.educause.edu

Blikstein P (2011) Using learning analytics to assess students' behavior in open-ended programming tasks. In: Proceedings of the 1st international conference on learning analytics and knowledge—LAK '11, p 110. https://doi.org/10.1145/2090116.2090132

Bowles N (2019, April 21) Silicon Valley Came to Kansas Schools. That Started a Rebellion. The New York Times. https://www.nytimes.com/2019/04/21/technology/silicon-valley-kansas-schools.html

Bowman DA, Coquillart S, Froehlich B, Hirose M, Kitamura Y, Kiyokawa K, Stuerzlinger W (2008) 3D user interfaces: new directions and perspectives. IEEE Comput Graph Appl 28(6):20–36. https://doi.org/10.1109/MCG.2008.109

Boyce S, O'Halloran J (2020) Active learning in computer-based college algebra. Primus 30(4):458–474. https://doi.org/10.1080/10511970.2019.1608487

Bradeško L, Mladenić D (2012) A survey of Chabot systems through a Loebner prize competition. In: Proceedings of Slovenian language technologies society eighth conference of language technologies, p 5

Case A (2018, June 22) Why gesture-based interfaces haven't lived up to the hype. Medium. https://medium.com/@caseorganic/why-gesture-based-interfaces-havent-lived-up-to-the-hype-9ab47aa3a94b

Cavanagh T, Chen T, Lahcen B, Paradiso J (2020) Constructing a design framework and pedagogical approach for adaptive learning in higher education: a practitioner's perspective. Int Rev Res Open Distrib Learn 21(1):172–196

Chaiklin S (2003) The zone of proximal development in Vygotsky's analysis of learning and instruction. In: Kozulin A, Gindis B, Ageyev VS, Miller SM (eds) Vygotsky's educational theory in cultural context. Cambridge University Press, pp 39–64. https://doi.org/10.1017/CBO9780511840975.004

Chen H, Ciborowska A, Damevski K (2019) Using automated prompts for student reflection on computer security concepts. In: Proceedings of the 2019 ACM conference on innovation and technology in computer science education—ITiCSE '19, pp 506–512. https://doi.org/10.1145/3304221.3319731

Chi M, Barnes T (2014) Educational data mining for individualized instruction in stem learning environments [NSF award database]. National Science Foundation Award Search. https://nsf.gov/awardsearch/showAward?AWD_ID=1432156

Chiang AC-C, Fung IP-W (2004) Redesigning chat forum for critical thinking in a problem-based learning environment. Internet High Educ 7(4):311–328. https://doi.org/10.1016/j.iheduc.2004.09.006

Ciolacu M, Tehrani AF, Beer R, Popp H (2017) Education 4.0—fostering student's performance with machine learning methods. In: 2017 IEEE 23rd international symposium for design and technology in electronic packaging (SIITME), pp 438–443. https://doi.org/10.1109/SIITME.2017.8259941

Cowie R (2014) Ethical issues in affective computing. Oxford University Press. https://doi.org/10.1093/oxfordhb/9780199942237.013.006

Dietrich D (2015, February) Why instructor satisfaction cannot be ignored. ELearn Magazine, an ACM Publication. https://elearnmag.acm.org/archive.cfm?aid=2735931

Digital Divide Compounds U.S. Education Equity Problem, First-of-Its-Kind Survey Reveals (2018, September 6) ACT Center for Equity in Learning. https://equityinlearning.act.org/press-releases/digital-divide-compounds-u-s-education-equity-problem-first-of-its-kind-survey-reveals/

Do L (2019) Study takes aim at biased AI facial-recognition technology. phys.org. https://phys.org/news/2019-02-aim-biased-ai-facial-recognition-technology.html

Duo S, Song LX (2012) An E-learning system based on affective computing. Phys Procedia 24:1893–1898. https://doi.org/10.1016/j.phpro.2012.02.278

EdX: About Us (2013, November 12) EdX. https://www.edx.org/about-us

Eicher B, Polepeddi L, Goel A (2018) Jill Watson Doesn't care if you're pregnant: grounding AI ethics in empirical studies. In: Proceedings of the 2018 AAAI/ACM conference on AI, ethics, and society—AIES '18, pp 88–94. https://doi.org/10.1145/3278721.3278760

Fatahi S, Moradian S (2018) An empirical study on the impact of using an adaptive e-learning environment based on learner's personality and emotion. p 8

Flipping the Classroom (2019) Center for Teaching and Learning. https://www.washington.edu/teaching/teaching-resources/engaging-students-in-learning/flipping-the-classroom/. Accessed 17 May 2019

Fonte FAM, Nistal ML, Rial JCB, Rodriguez MC (2016) NLAST: a natural language assistant for students. In: 2016 IEEE global engineering education conference (EDUCON), pp 709–713. https://doi.org/10.1109/EDUCON.2016.7474628

Fuad M, Akbar M, Zubov L (2018a) Active learning for out-of-class activities by using interactive mobile apps. In: Sixth international conference on learning and teaching in computing and engineering. https://par.nsf.gov/biblio/10057677-active-learning-out-class-activities-using-interactive-mobile-apps

Fuad M, Akbar M, Zubov L (2018b) Dysgu: a mobile-based adaptive system to redesign out-of-class activities. In: 2018 IEEE frontiers in education conference (FIE), pp 1–5. https://doi.org/10.1109/FIE.2018.8659143

García P, Amandi A, Schiaffino S, Campo M (2007) Evaluating Bayesian networks' precision for detecting students' learning styles. Comput Educ 49(1):794–808. https://doi.org/10.1016/j.compedu.2005.11.017

Ghali R, Ouellet S, Frasson C (2016) LewiSpace: an exploratory study with a machine learning model in an educational game. J Educ Train Stud 4(1):192–201

Goshevski D, Veljanoska J, Hatziapostolou T (2017) A review of gamification platforms for higher education. In: Proceedings of the 8th Balkan conference in informatics on BCI '17, pp 1–6. https://doi.org/10.1145/3136273.3136299

Haba E (2017, September 14) M-Shule: artificial intelligence for personalized learning. Engineers Without Borders Canada Latest News. https://www.ewb.ca/en/news-and-events/news/m-shule-artificial-intelligence-personalized-learning/

Harackiewicz JM, Sansone C, Blair LW, Epstein JA, Manderlink G (1987) Attributional processes in behavior change and maintenance: smoking cessation and continued abstinence. J Consult Clin Psychol 55(3):372–378. https://doi.org/10.1037/0022-006X.55.3.372

IBM (2018, November 16) Watson Assistant | IBM Cloud. Watson Assistant | IBM Cloud. https://www.ibm.com/cloud/watson-assistant/

Isomöttönen V, Lakanen A-J, Lappalainen V (2019) Less is more! Preliminary evaluation of multi-functional document-based online learning environment. In: 2019 IEEE frontiers in education conference (FIE), pp 1–5. https://doi.org/10.1109/FIE43999.2019.9028353

Knight W (2017) The Dark Secret at the Heart of AI; No one really knows how the most advanced algorithms do what they do. That could be a problem. MIT Technology Review—Artificial Intelligence/Machine Learning. https://www.technologyreview.com/s/604087/the-dark-secret-at-the-heart-of-ai/

Kumar AN (2017) The effect of providing motivational support in parsons puzzle tutors. In André E, Baker R, Hu X, Ma, Rodrigo MT, du Boulay B (eds), Artificial intelligence in education, vol 10331. Springer International Publishing, pp 528–531. https://doi.org/10.1007/978-3-319-61425-0_56

Kumar AN (2018) Epplets: a tool for solving parsons puzzles. In: Proceedings of the 49th ACM technical symposium on computer science education—SIGCSE '18, pp 527–532. https://doi.org/10.1145/3159450.3159576

Kyrilov A (2014) Using case-based reasoning to improve the quality of feedback generated by automated grading systems. In: Proceedings of the tenth annual conference on international computing education research—ICER '14, pp 157–158. https://doi.org/10.1145/2632320.2632330

Kyrilov A, Noelle DC (2015) Binary instant feedback on programming exercises can reduce student engagement and promote cheating. In: Proceedings of the 15th Koli Calling conference on computing education research—Koli Calling '15, pp 122–126. https://doi.org/10.1145/2828959.2828968

Kyrilov A, Noelle DC (2016) Do students need detailed feedback on programming exercises and can automated assessment systems provide it? J Comput Sci Coll 34(1):7

Lateef F (2010) Simulation-based learning: just like the real thing. J Emerg Trauma Shock 3(4):348–352. https://doi.org/10.4103/0974-2700.70743

Lehnert WG (1977) The process of question answering. PhD Thesis, Yale University

Li FWB, Rynson WHL, Dharmendran P (2010) An adaptive course generation framework. Int J Dist Educ Technol 8(3):47–64. https://doi.org/10.4018/jdet.2010070104

Lu JJ, Harris LA (2018) Artificial Intelligence (AI) and Education (Congressional Report No. IF10937; IN FOCUS). Library of Congress. https://fas.org/sgp/crs/misc/IF10937.pdf

Lynch M (2016, August 23) Focus on these four areas to create a classroom environment conducive to learning. The Edvocate. https://www.theedadvocate.org/focus-four-areas-create-classroom-environment-conducive-to-learning/

Ma W, Adesope OO, Nesbit JC, Liu Q (2014) Intelligent tutoring systems and learning outcomes: a meta-analysis. J Educ Psychol 106(5):901–918. https://doi.org/10.1037/a0037123

Marr B (2018, July 25) How is AI used in education—real world examples of today and a peek into the future. How is AI used in education. https://www.forbes.com/sites/bernard-marr/2018/07/25/how-is-ai-used-in-education-real-world-examples-of-today-and-a-peek-into-the-future/#79ea483b586e

Martinez R, Hong H, Lee D-W (2013) Automated essay scoring system by using support vector machine. Int J Adv Comput Technol 5:316–322. https://doi.org/10.4156/ijact.vol5.issue11.37

Miller R (2017, August 26) AI will fundamentally change how we manage content. TechCrunch. http://social.techcrunch.com/2017/08/26/ai-will-fundamentally-change-how-we-manage-content/

Mostafavi B, Barnes T (2017) Evolution of an intelligent deductive logic tutor using data-driven elements. Int J Artif Intell Educ 27(1):5–36. https://doi.org/10.1007/s40593-016-0112-1

National Research Council, Division of Behavioral and Social Sciences and Education, Center for Education, Board on Testing and Assessment, & Committee on the Foundations of Assessment (2001) Knowing what students know: the science and design of educational assessment. National Academies Press

Nehm RH, Ha M, Mayfield E (2012) Transforming biology assessment with machine learning: automated scoring of written evolutionary explanations. J Sci Educ Technol 21(1):183–196. https://doi.org/10.1007/s10956-011-9300-9

Nsehe M (2019, February 25) Meet The 10 African Startups Competing For The "Next Billion" EdTech Prize In Dubai. Forbes. https://www.forbes.com/sites/mfonobongnsehe/2019/02/25/meet-the-10-african-startups-competing-for-the-next-billion-edtech-prize-in-dubai/

Official Site | Second Life—Virtual Worlds, Virtual Reality, VR, Avatars, Free 3D Chat (2019). https://secondlife.com/. Accessed 28 Apr 2019

Ogor EN (2007) Student academic performance monitoring and evaluation using data mining techniques. In: Electronics, robotics and automotive mechanics conference (CERMA 2007), pp 354–359. https://doi.org/10.1109/CERMA.2007.4367712

Ontañón S, Valls-Vargas J, Zhu J, Smith BK, Char B, Freed E, Furqan A, Howard M, Nguyen A, Patterson J (2017) Designing visual metaphors for an educational game for parallel programming. In: Proceedings of the 2017 CHI conference extended abstracts on human factors in computing systems—CHI EA '17, pp 2818–2824. https://doi.org/10.1145/3027063.3053253

OpenSimulator (2019). http://opensimulator.org/wiki/Main_Page. Accessed 28 Apr 2019

OPNET Optimum Network Performance (2020) OPNET Network Simulator. Opnet Network Simulator. http://opnetprojects.com/opnet-network-simulator/. Accessed 12 June 2020

Overview of ALEKS (2020). https://www.aleks.com/about_aleks/overview. Accessed 22 May 2020

Paaßen B, Hammer B, Price TW, Barnes T, Gross S, Pinkwart N (2018) The continuous hint factory—providing hints in vast and sparsely populated edit distance spaces. J Educ Data Min 10(1):1–35. https://doi.org/10.5281/zenodo.3554698

Peddycord-Liu Z, Mostafavi B, Barnes T (2016) Combining worked examples and problem solving in a data-driven logic tutor. In: Micarelli A, Stamper J, Panourgia K (eds) Intelligent tutoring systems. Springer International Publishing, pp 347–353

Petersen SE, Ostendorf M (2009) A machine learning approach to reading level assessment. Comput Speech Lang 23(1):89–106. https://doi.org/10.1016/j.csl.2008.04.003

Petkovic D, Okada K, Sosnick M, Iyer A, Zhu S, Todtenhoefer R, Huang S (2012) Work in progress: a machine learning approach for assessment and prediction of teamwork effectiveness in software engineering education. In: 2012 frontiers in education conference proceedings, pp 1–3. https://doi.org/10.1109/FIE.2012.6462205

Picard RW (2003) Affective computing: challenges. Int J Hum Comput Stud 59(1–2):55–64. https://doi.org/10.1016/S1071-5819(03)00052-1

Polachowska K (2019, June 13) AI in education: can AI improve the way we teach and learn?—Neoteric. Software House That Helps You Innovate—Neoteric. https://neoteric.eu/blog/ai-in-education-can-ai-improve-the-way-we-teach-and-learn/

Price TW, Dong Y, Barnes T (2016) Generating data-driven hints for open-ended programming. In: 9th international conference on educational data mining, pp 191–198

Reyna J (2011) Digital teaching and learning ecosystem (DTLE): a theoretical approach for online learning environments. Proc Ascilite 2011:1083–1088

Rezaei MS, Montazer GA (2016) An automatic adaptive grouping of learners in an e-learning environment based on fuzzy grafting and snap-drift clustering. Int J Technol Enhanc Learn 8(2):169. https://doi.org/10.1504/IJTEL.2016.078090

Schmelzer R (2019, July 12) AI applications in education. Forbes https://www.forbes.com/sites/cognitiveworld/2019/07/12/ai-applications-in-education/

Shermis MD, Burstein J, Higgins D, Zechner K (2010) Automated essay scoring: writing assessment and instruction. In: International encyclopedia of education. Elsevier, pp 20–26. https://doi.org/10.1016/B978-0-08-044894-7.00233-5

Spector C (2017, December 5) Students' early test scores don't predict academic growth over time. Stanford News. https://news.stanford.edu/2017/12/05/students-early-test-scores-not-predict-academic-growth-time/

Srinivasan D (2018, November 7) Powered exoskeletons are the technology of the future. Silicon Republic. https://www.siliconrepublic.com/machines/divya-srinivasan-exoskeletons-virginia-tech

Strauss J (2017) The surprising thing Google learned about its employees—and what it means for today's students. The Washington Post Answer Sheet—Analysis. https://www.washingtonpost.com/news/answer-sheet/wp/2017/12/20/the-surprising-thing-google-learned-about-its-employees-and-what-it-means-for-todays-students/?noredirect=on&utm_term=.215212d4f366

Summit Learning Platform (2017, April 14) Chan Zuckerberg Initiative. https://chanzuckerberg.com/newsroom/summit-learning-platform/

The Complete Guide to Virtual Labs (2020) Labster. https://www.labster.com/the-complete-guide-to-virtual-labs/. Accessed 21 May 2020

Tsinakos AA (2006) Virtual instructor and pedagogical issues. In: Sixth IEEE international conference on advanced learning technologies (ICALT '06), pp 1123–1124. https://doi.org/10.1109/ICALT.2006.1652654

US EPA (2013, September 25) Section 508 Standards [Policies and Guidance]. Section 508: Accessibility | US EPA. https://www.epa.gov/accessibility/section-508-standards

VanLehn K, Burkhardt H, Cheema S, Kang S, Pead D, Schoenfeld A, Wetzel J (2019a) Can an orchestration system increase collaborative, productive struggle in teaching-by-eliciting classrooms? Interact Learn Environ 0(0):1–19. https://doi.org/10.1080/10494820.2019.1616567

VanLehn K, Cheema S, Kang S, Wetzel J (2019b) Auto-sending messages in an intelligent orchestration system: a pilot study. In: Isotani S, Millán E, Ogan A, Hastings P, McLaren B, Luckin R (eds) Artificial intelligence in education. Springer International Publishing, pp 292–297. https://doi.org/10.1007/978-3-030-23207-8_54

Virtual Science Labs at Your Fingertips (2020). https://praxilabs.com/en/virtual-labs. Accessed 21 May 2020

Vujičić T, Simonović D, Đukić A, Šestić M (2018) Browninfo methodology and software for development of interactive brownfield databases. In: Hadžikadić M, Avdaković S (eds) Advanced technologies, systems, and applications II. Springer International Publishing, pp 484–502

Wadsworth BJ (1996) Piaget's theory of cognitive and affective development: foundations of constructivism, 5th edn. Longman Publishing

Waltz E (2019, February 8) Hey, Siri: read my lips. IEEE Spectrum: Technology, Engineering, and Science News

Wang K, Singh R, Su Z (2018) Search, align, and repair: data-driven feedback generation for introductory programming exercises. In: Proceedings of the 39th ACM SIGPLAN conference on programming language design and implementation, pp 481–495. https://doi.org/10.1145/3192366.3192384

Wilensky R (1977) PAM: a program that infers intentions. In: Proceedings of the 5th international joint conference on artificial intelligence, vol 1, pp 15–15. http://dl.acm.org/citation.cfm?id=1624435.1624438

Wilensky R, Chin DN, Luria M, Martin J, Mayfield J, Wu D (2000) The Berkeley UNIX consultant project. Artif Intell Rev 14(1–2):43–88. https://doi.org/10.1023/A:1006500224529

Winkelmann K, Keeney-Kennicutt W, Fowler D, Macik M (2017) Development, implementation, and assessment of general chemistry lab experiments performed in the virtual world of second life. J Chem Educ 94(7):849–858. https://doi.org/10.1021/acs.jchemed.6b00733

Winston PH (2016) The genesis story understanding and story-telling system—a 21st century step toward artificial intelligence. http://groups.csail.mit.edu/genesis/papers/StoryWhitePaper.pdf

Wiratunga N, Adeyanju I, Coghill P, Pera C (2011) RubricAce TM: a case-based feedback recommender for coursework assessment★. In: *Proceedings of 16th UK workshop on case-based reasoning*, p 11. http://ceur-ws.org/Vol-829/paper3.pdf

Wu C-H, Huang Y-M, Hwang J-P (2016) Review of affective computing in education/learning: trends and challenges: advancements and trends of affective computing research. Br J Educ Technol 47(6):1304–1323. https://doi.org/10.1111/bjet.12324

Zhou G, Lynch CF, Price TW, Barnes T, Chi M (2016) The impact of granularity on the effectiveness of students' pedagogical decision. Cogn Sci 2016:2801–2806. https://cogsci.mindmodeling.org/2016/papers/0482/index.html

Emergent Guiding Principles for STEM Education

Lawrence C. Ragan and Lorraine J. Ramirez Villarin

1 Motivation

STEM-related jobs are on the rise. Workers are required to exercise critical thinking and decision-making skills while being knowledgeable and competent in domains related to Science, Technology, Engineering, and Math (STEM). The eXploring the Future of Innovative Learning Environments (X-FILEs) Workshop, hosted by the Florida Institute of Technology in November of 2018, allowed stakeholders to make informed decisions about the adoption and use of innovative learning environments (ILEs) in higher STEM education. Participants had the opportunity to consider four emerging technologies that could assist in this effort: personalized and adaptive learning, multimodal learning formats, cross reality (XR), and artificial intelligence and machine learning. The research team gathered shared ideas through online meetings and collaborative activities that reflect on the opportunities and challenges to expect while implementing ILEs in higher education STEM curricula.

Since these immersive technologies are quickly evolving, it has been challenging for institutes to implement integrated STEM education programs that utilize them (Blackley and Howell 2015; Briener et al. 2012). Therefore, these guiding principles are a good starting point.

L. C. Ragan (✉)
Ragan Education, State College, PA, USA
e-mail: lcr1@psu.edu

L. J. Ramirez Villarin
University of North Georgia, Dahlonega, GA, USA
e-mail: Lorraine.RamirezVillarin@ung.edu

© The Author(s) 2021
J. Ryoo, K. Winkelmann (eds.), *Innovative Learning Environments in STEM Higher Education*, SpringerBriefs in Statistics,
https://doi.org/10.1007/978-3-030-58948-6_6

2   Recurring Themes

A series of recurring themes emerged from the data collected at the 2018 workshop. These themes appeared across most or all four technology categories and served as the basis for the development of guiding principles for the future of STEM education. In most, if not all cases, the identified themes were interconnected and related to each other. For example, providing students a personalized learning environment based on personal interest and preferences is dependent upon having collected data about each student. In collecting data on student preferences, concern for student rights to privacy must be evaluated.

The nine emerging principles were classified into four clusters to better understand their correlation and importance within the academic instructional design field. The four clusters encompass the student experience, the learning community, its availability, and the learners' protection. While considering the affordances of the supporting technologies in the generation of ILEs, the limitations of these technologies must also be weighed.

2.1   Student Experience

Student-Engaged Learning Environment While the lecture format of instruction has been the norm in higher education for years, it is time for the students to play a more active role in their educational experience. STEM education requires learners to solve real-world challenges through the application of scientific and mathematical principles, the engineering process, and the use of current and emerging technologies. To help retain learners who choose STEM programs in college, active learning approaches are a must (Olson and Riordan 2012). Carefully crafted ILEs allow students to construct representations of their own knowledge and practice applying their skills to new and novel problem scenarios. Donally (2018) suggests students create monuments for a history project, design worlds with unique ecosystems for science, or bring a literary reference to their reality for a reading assignment.

Personalized learning provides the opportunity to tailor instruction to the individual's needs, skills, goals, and learning preferences while consistently monitoring progress (Sampson 2001, cited in Sampson et al. 2002). This approach promotes improvement while enabling participants to take control of their learning. Nonetheless, student-centered environments like multimodal learning require students to have a high degree of agency and metacognition to recognize when learning is happening and when to challenge themselves (Bezemer and Kress 2016; Moreno and Mayer 2007; Phuong et al. 2017; Sankey et al. 2010).

Virtual reality provides an opportunity to keep the student as the center of the learning environment. While being "mistake-free" (Bailenson 2018), participants actively explore virtual worlds or encounter scenarios where learning takes place through trial and error. Cross reality experiences like these have shown equivalent learning gains (Madden et al. 2018; Winkelmann et al. 2017) or lesser learning gains than traditional instruction (Makransky et al. 2019). However, cross reality experiences focus more on scientific thinking processes and behaviors than it does in content knowledge.

Applications of artificial intelligence help craft these learning environments through automated tutoring, personalized learning, student knowledge assessment, and automating mundane instructors' tasks (Lu and Harris 2018). Devices like chatbots and auto-graders enable progress monitoring which may pinpoint strengths and weaknesses that can steer students and instructors on the proper academic path.

Personalized Learning Students do not learn in the same way. They learn better when instruction is individualized to the learner's needs (Kerr 2016). However, factors like academic background, culture, language, or disability may influence the manner an individual learns. Having the advantage of a personalized or customized learning environment can be beneficial for students, as well as instructors. Dashboards that monitor individual activity, provide feedback, alert instructors of students' difficulties, and suggest other approaches are likely to enrich instruction. For example, adaptive learning environments are based on individual students' profiles created upon their strengths, weaknesses, and pace of learning (Becker et al. 2018). The technology tracks progress and adjusts the learning path enabling instructors to deliver timely feedback. These adaptive learning systems promote self-regulated learning which has been shown to be a predictor of academic success (Boekaerts 1997; Miltiadou and Savenye 2003; Pintrich and De Groot 1990; Rosen et al. 2010). A study by Prain et al. (2013) reported that math students participating in a personalized curriculum got better scores and were more engaged than those who were not.

A "one size fits all" curriculum may not fulfill everyone's needs (Gee 1996; Phuong et al. 2017). While studying augmented reality applications for cross reality, one recommendation made by Radu (2014) was that experiences be designed based on curriculum and pedagogical needs while providing the opportunity for faculty to customize them accordingly. For example, an AI-driven personalized learning system would adjust the context of the content based on the learner's interest and preferred learning modality while still meeting desired learning outcomes.

In adaptive learning systems, students work at their own pace. Since learner activities are auto-graded, they can receive feedback and scaffolding when needed. It is important that the student does not feel isolated. Communication should not be a concern between students and instructors. As in any learning experience, support should be available at any time in the form of live instructors or automated systems to provide guidance or answer questions.

Flexible, Fluid, and Evolutionary Current and emerging technologies are transforming constantly tending to learners' and instructors' needs while also considering academic institutions' demands. The required technology for most of these ILEs can already be found in many homes, classrooms, and workplaces in the form of smartphones, computers, or game systems. Bailenson (2018: 9) predicts virtual reality to be "a mainstream technology, worth an estimated $60 billion, in the next decade." As for augmented reality, it is expected to reach $60 billion in 2020 (Porter and Heppelmann 2017). This flexibility of access provides the opportunity for informal learning environments almost anywhere. Multiuser 3D virtual world environments allow geographically dispersed learners to participate in a traditional classroom-like environment while learning at their own pace (Olasoji and Henderson-Begg 2010) and in their own time zone. This provides instructors the opportunity to convey fluid and seamless learning experiences inside and outside the classroom.

The flexibility of content and context is another advantage of ILEs. Having the malleability to enhance content by presenting it in more than one sensory mode or incorporating text-based information and multimedia using a diversity of modes can generate multiple access points for learning (Bezemer and Kress 2016; Matusiak 2013; Nouri 2018; Sankey et al. 2010) as those provided through multimodal formats. Well-designed courses like those for adaptive learning foster the learning of processes and strategies that promote self-regulation in students (Dabbagh and Kitsantas 2013; Hense and Mandl 2012; Shea and Bidjerano 2009). On the other hand, higher interactive virtual environments work through learning activities as immersion stimulates emotional involvement which is essential in the relocation of learning to long-term memory (Aldrich 2009).

For measuring student progress and performance, formative and summative assessments inform the student of advancement toward the learning objectives and provide a final grade of achievement. However, many methods do not objectively address student achievement or translate into real-world outcomes in STEM fields (Spector 2017). Using "real-to-life" and situated learning can enable students to practice or be measured within the context of the domain. Simulations, virtual reality, and augmented reality can recreate "real-world" experiential opportunities for learning that would otherwise be impossible in a regular classroom. These immersive environments promote learning without any hazards to students or others. For example, surgical procedures may be practiced safely as no "real" patients will be in danger (Health Scholars n.d.). Chemicals can be mistakenly combined or spilled and cleaned up by pressing a recycle button (Faulconer and Gruss 2018) with no harm to participants. These experiences can be paused, reset, and available on-demand (Lynch and Ghergulescu 2017), while the dynamic nature of the computer system records and gathers data in the background as the learner moves along (Rose 1995). Learner activities like these bring relevance to context while replicating many tasks professionally encountered in STEM-related fields.

STEM educators need to be open to a variety of technologies and recognize the interoperability of various systems. Learning management systems (LMS) may not provide the instructor with all they need to support a synchronous or asynchronous learning system. In some cases, there are add-ons to the LMS that may be used to expand the system's functionality. Social media platforms, cross-platform messaging applications, and other application software may facilitate instruction even if these programs' main duty is not related to academics. For example, Twitter can be used to promote interaction among peers and instructors, while Facebook and WhatsApp are flexible enough to create groups or chats for which you can share video and audio recordings and get immediate feedback.

2.2 Learning Community

Include Both Individual and Group Interactions and Input Students who select STEM programs in college might be looking forward to engaging in active learning approaches. To challenge these learners, instructors may want to alternate between some of these technologies for individual and collaborative tasks. For example, computer-supported collaborative work assesses individual and collaborative potentials based on individual and shared contributions. Ward and Sonneborn (2009) indicate that creative problem-solving in groups must consider the learner group contribution and the quality and quantity of ideas produced through individual collaborations. Other communication systems like collaborative virtual environments allow users to share a three-dimensional digital space while occupying remote physical locations (Yee and Bailenson 2006). With dozens of multiuser experiences available (JuegoStudios 2019) like social virtual reality instruction, facilitated discussion, and project collaboration, instructors and researchers have the freedom to incorporate these for either learning or research purposes.

As for personalized and adaptive learning, interactions occur between student and content and between student and instructor. Student-student interactions are not typical. Moore (1989: 1) describes student-content interaction as that which "results in changes in the learner's understanding, the learner's perspective, or the cognitive structures of the learner's mind." Student-instructor interaction, often based on performance, usually takes place through the learning management system.

Besides promoting collaborative learning, interactions that can foster communication skills and cultural enrichment could enhance these learning experiences. Donally (2018) claims that augmented reality, virtual reality, and mixed reality allow students to shape how they view others around the world through interaction and collaboration. It should be a top priority for institutions with access to these cross reality resources to have the opportunity to stimulate interaction and promote collaboration while inviting students from other academic institutions with similar access. These immersive technologies should expand to engage global interactions and feedback.

Multimodal learning environments facilitate interaction since those who might have trouble communicating in one mode can interact in another. For example, students can arrange to connect with culturally diverse students who might be well-versed in the digital world or other multilingual or English language learners or others who have similar learning disabilities, creating a rich, comfortable learning experience. Considering the evolving demographics of students expected in higher education, multimodal formats can help customize student-centered experiences and address diversity issues.

Co-contributed and Multiply Sourced Increasingly, with access to sophisticated online content development tools, the instructional material used in a course may be generated from multiple sources. Students, generating content through social media or self-publishing, perceive a class assignment as an opportunity to create, generate new information, or assimilate existing data into new perspectives. This student-generated content makes an increasingly critical contribution to the course experience. Indeed, with the trend toward active and engaged learning, students are being encouraged and directed to create new works, products, and ideas that may have added value outside of the course.

The content for today's classes is unlikely to be singularly sourced, that is, comes from a single text or course pack. Content from multiple sources can be presented in different modes (e.g., gestures, visuals, multimedia, text-based information) (Bezemer and Kress 2016; Matusiak 2013; Nouri 2018; Sankey et al. 2010) and greatly enhance the learning experience. The richness of content digitally accessible is extensive and more likely to be integrated into the learning materials than other reference sources. In many cases, live data streams provide the most current data source. For example, assembling a virtual expert panel of industry leaders directly linked into a live lecture can further approximate reality and enrich the experience for all participants. With this access to a plethora of content comes the responsibility to ensure the accuracy and validity of the information. This alone becomes a new skill the lifelong learner will need to master!

2.3   Availability

Equity of Access As new learning systems emerge, the concern of equitable access for all learners needs to be reviewed. These ILEs are crucial learning instruments, and all students should be provided equal access regardless of privilege, ability, or economic situation. More sophisticated adaptive learning systems, often developed commercially, may also increase the cost for engagement for the learner. Even as the cost of cross reality devices may decrease, the technical requirements may include higher-speed bandwidth, device batteries, and add-on components. In some locations, for example, broadband access may limit student interaction with cross reality systems.

Technologies such as artificial intelligence can require resources from machine learning models for each assessment and grade level (Balfour 2013; Shermis et al. 2010; Nehm et al. 2012). Curating learning environments for different modes can also be quite complex and demand additional allocation of resources (Bezemer and Kress 2016). Add to this finding the preparation of the physical space including enhanced lighting and audio, upkeep of devices, and the time and money it takes to train instructors and technicians. This will make it challenging for certain academic institutions who lack the resources. A digital divide still exists between certain educational institutions and communities (ACT Center for Equity in Learning 2018).

Accessibility Assistive technology refers to any item, piece of equipment, software program, or product that increases, maintains, or improves the functional capabilities of people with disabilities and is in the form of equipment, software, or hardware (Assistive Technology Industry Association 2019). Immersive learning environments such as those made possible through virtual worlds simulations can provide participants with a sense of physical movement which can enhance their experience.

This same advantage, however, in some circumstances, may deprive an individual with disabilities from this enhanced experience (Ryan 2019). The Americans with Disabilities Act prohibits discrimination toward this population by allowing support in the workplace and academic institutions (U.S. Equal Employment Opportunity Commission n.d.; U.S. General Services Administration 2018). Unfortunately, most personal devices, nowadays, do not accommodate the emerging technological demands for replicating experiences for this group and other diverse learners. Therefore, engagement will be interrupted as you exit the classroom. Providing this population with the same opportunities as traditional student has is not only keeping compliant with the law but also the right thing to do to achieve equity and inclusiveness.

2.4 Learner Protection

Safe and Secure Keeping the privacy of users seemed to be a recurring concern across three of the four technologies explored. Unfortunately, the digital world is not exempt from inappropriate or unlawful actions. Users may be exposed to threats like cyberbullying, griefing, and security attacks. For example, research done by Outlaw and Duckles (2017) reflected that female participants in social virtual reality experiences had encountered "flirting, a lack of respect for personal boundaries, and socially undesirable behavior," similar to those in real life, leaving little desire to return to these platforms. As for multiuser environments, the number of accounts and avatars that can be created can increase trust and identity concerns (Warburton 2009). In order to minimize this, proper training would be required for instructors and learners since a lack of technical skills could make them more susceptible to this kind of action. Faculty and students need to understand that some channels or modes may be "open" for public viewing. Professional development can provide orientation on what university tools are available to ensure students' safety and privacy. Nevertheless, training is time-consuming and costly.

For example, during the COVID-19 pandemic, Zoom became one of the most popular platforms for teleconferencing. However, "uninvited guests" would ruin the video conference experience using shocking imagery, racial slurs, or profanity leading to a phenomenon known as Zoom bombing (Lorenz and Alba 2020). These "trolls" (disruptive, uninvited strangers) were hard to identify and, therefore, suffered little to no consequences for their malicious actions. Zoom implemented stricter measures of access control enabling the host to have more restrictions over the meeting. Nonetheless, these incidents may still happen in other platforms and without the user's knowledge. For example, spyware can be installed remotely through an email, a photo download, or an instant message. This type of malware records screenshots of information typed (e.g., passwords, usernames), and media accessed.

Access control measures can be a challenge, but there are certain options that can improve security. More and more organizations are adopting multifactor authentication which requires users to use more than one physical device to verify their identity. The second factor device can be your mobile phone in addition to a personal computer. Facial recognition and fingerprint readers are also common biometrics mechanisms used for identity authentication. Other options include the power to access your information from another device which gives you the flexibility to receive notifications when changes are made in your accounts. To protect your information, data encryption allows the system to scramble the data and share a secret key or key pair with those who access it. However, as technology changes every day, access control measures need to evolve to outsmart attackers.

Ethics The increasing technical dimensions of developing STEM curriculum demands a critical review of the ethical aspects of the use of these tools in future educational systems. As presented previously, the safety and privacy of each learner are paramount. Although proper netiquette is expected among users, hacking and cyberbullying are dangers to which everyone is exposed. As users create multiple accounts and avatars, for example, in a multiuser environment, trust and identity may be threatened (Warburton 2009). These hazards can be intimidating for some and may limit use especially if an individual is not familiar with the different technologies. To minimize falling into these negative experiences, rigorous training on the different modes and technologies is a must for instructors and learners going through this digital transition. However, this may not be enough. Certain corporations, like Microsoft, are adding face-to-face interactions to virtual and persistent collaboration rooms to enhance trust as they hope to expand to the workforce market with a cross reality platform called Hololens (Weise 2019).

As technologies involving cross reality evolve bringing features like sight, sound, smell, taste, pressure, heat, and texture to the learning environment, the danger exists in disconnecting the user from reality. Bailenson (2018: 46) indicates that "VR feels real, and its effects on us resemble the effects of real experiences. Consequently, a VR experience is often better understood *not as a media experience, but as an actual experience,* with the attendant results for our behavior." If virtual experiences are perceived as real, for example, it raises the question of ethics in conducting research with "virtually real-life" specimens and "real-life" specimens.

Another ethical challenge relates to the way algorithms work in predicting performance. Learners might be presented with opportunities or support based on their anticipated scores or scores provided by a machine algorithm (Ogor 2007; Ciolacu et al. 2017). Users may want to "play the system" to deceitfully display to have a better performance. Instructors may feel uncomfortable trusting results from an algorithm and may resort to other forms of assessment. However, FERPA requires student information to be kept private, including scores. To keep compliant, strict rules of engagement, regulations, and set boundaries should be established for users and offenders.

The use of increasingly intrusive technologies that collect, store, analyze, and manipulate personal data presents a myriad of ethical decisions that must be addressed and defined. Although AI and machine learning has the capacity to track and adjust the course content to the interest and preferences of the learner, they also present the opportunity to negatively influence students' behaviors. Defining the boundaries of appropriate social behavior when presented such powerful data gathering tools is a crucial step in the development of future learning systems.

3   Conclusion

The guiding principles for STEM education emerging from the X-FILEs Workshop provide a foundation on which to consider the design and development of future STEM innovative learning environments. These principles are not exclusive to STEM education but are also critical to the construction of all course instruction in the near and future education systems. These guiding principles reflect the capability and promise of the technologies that make it possible as well as the challenges and threats presented by the use of these technologies.

The guiding principles emerging from the X-FILEs Workshop aggregated into four clusters representing the student experience, the learning community, availability, and learner protection. All of the clusters reflect some aspect of the learner's response to the design of the learning system. Many of the principles are interrelated and, in some cases, represent "the other side of the coin" of a stated principle. For example, a personalized learning system described in Sect. 2.1 also presents the challenges of protecting personal data gathered from profile information of the student as referenced in Sect. 2.4.

Forefront in these principles is a focus on the changing role and experience of the learner. These principles embrace an engaged and active learner as both a recipient and a contributor to the learning system. Capitalizing on the capability for all class participants to contribute to the construction of the course content, the principle stated in Sect. 2.2, empowers students to improve and enhance the course materials. In increasing the responsibility of the learner as a co-contributor, the act of generating, analyzing, and publishing course content also prepares them for controlling and directing their own future learning opportunities.

Perhaps the most challenging ideas and unanswered questions evolve from the ideas in the Sect. 2.4 principle. The promise and power of the previous principles raise the awareness and concern for addressing how to protect the individual and group rights of learners. The unprecedented affordances and opportunities enabled by the technologies and methodologies explored in the X-FILEs research create a multitude of ethical questions regarding harvesting, analyzing, and manipulating student data and input. Where, when, and how to use student data within the learning system need to be defined and appropriate guidelines and protections established.

The guiding principles emerging from the X-FILEs Workshop can serve as the basis for the design and development of student-centered, interactive, and increasingly effective innovative learning environments. These principles need to be further vetted, refined, and improved in order to address all dimensions of the learning ecosystem. Input and responses from all stakeholders also need to be garnered in order to ensure as thorough and comprehensive of a set of guidelines as possible. The future of STEM education is exciting and creates the responsibility of doing so thoughtfully and intentionally in order to maximize learning and preparing students for lifelong learning.

References

ACT Center for Equity in Learning (2018) Digital divide compounds U.S. education equity problem, first-of-its-kind survey reveals. 6 Sep 2018

Aldrich C (2009) Learning online with games, simulations, and virtual worlds. Jossey-Bass, San Francisco

Assistive Technology Industry Association (2019) What is AT? https://www.atia.org/at-resources/what-is-at/

Bailenson J (2018) Experience on demand: what virtual reality is, how it works, and what it can do. W. W. Norton & Company, New York

Balfour SP (2013) Assessing writing in MOOCs: automated essay scoring and calibrated peer review™. Res Pract Assess 8:40–48

Becker SA, Brown M, Dahlstrom E, Davis A, DePaul K, Diaz V, Pomerantz J (2018) NMC horizon report: 2018 higher education edition. EDUCAUSE, Louisville

Bezemer J, Kress G (2016) Multimodality, learning and communication: a social semiotic frame. Routledge, New York

Blackley S, Howell J (2015) A STEM narrative: 15 years in the making. Aust J Teach Educ 40(7). https://doi.org/10.14221/ajte.2015v40n7.8

Boekaerts M (1997) Self-regulated learning: a new concept embraced by researchers, policy makers, educators, teachers, and students. Learn Instr 7(2):161–186

Briener JM, Harkness SS, Johnson CC, Koehler CM (2012) What is STEM? A discussion about conceptions of STEM in education and partnerships. Sch Sci Math 112(1):3–11

Ciolacu M, Tehrani AF, Beer R, Popp H (2017, October) Education 4.0—fostering student's performance with machine learning methods. In: 2017 IEEE 23rd international symposium for design and technology in electronic packaging (SIITME). IEEE, pp 438–443

Dabbagh N, Kitsantas A (2013) Using learning management systems as metacognitive tools to support self-regulation in higher education contexts international handbook of metacognition and learning technologies. Springer, pp 197–211

Donally J (2018) Learning transported. Augmented, virtual, and mixed reality for all classrooms. International Society for Technology in Education, Portland

Faulconer E, Gruss A (2018) A review to weigh the pros and cons of online, remote, and distance science laboratory experiences. Int Rev Res Open Distrib Learn 19(2). https://doi.org/10.19173/irrodl.v19i2.3386

Gee JP (1996) Discourses and literacies. In: Luke A (ed) Social linguistics and literacies: ideology in discourses, 2nd edn. Taylor & Francis, London, pp 122–148

Health Scholars (n.d.) Fire in the ORTM Virtual Reality Simulation | Medical Training For Surgical Fires. https://www.youtube.com/watch?v=10Ke4kDSpGM&feature=youtu.be

Hense J, Mandl H (2012) Learning "in" or "with" games? Quality criteria for digital learning games from the perspectives of learning, emotion, and motivation theory. International Association for Development of the Information Society

JuegoStudios (2019, January 30) Popular various social VR platforms [Infographic]. https://www.juegostudio.com/infographic/various-social-vr-platforms

Kerr P (2016) Adaptive learning. ELT J 70(1):88–93. https://doi.org/10.1093/elt/ccv055

Lorenz T, & Alba D (2020) 'Zoombombing' becomes a dangerous organized effort. The New York Times, 3

Lu JJ, Harris LA (2018, August) Artificial Intelligence (AI) and Education. Library of Congress, Congressional Report IF10937

Lynch T, Gherghulescu I (2017, March) Review of virtual labs as the emerging technologies for teaching STEM subjects. In: INTED2017 Proc. 11th Int. Technol. Educ. Dev. Conf., Valencia Spain, 6–8 March, pp 6082–6091

Madden J, Won A, Schuldt J, Kim B, Pandita S, Sun Y, Stone T, Holmes N (2018, August 1–2) Virtual reality as a teaching tool for moon phases and beyond. Paper presented at Physics Education Research Conference 2018, Washington, DC. https://www.compadre.org/Repository/document/ServeFile.cfm?ID=14819&DocID=4966. Accessed 17 May 2019

Makransky G, Terkildsen TS, Mayer RE (2019) Adding immersive virtual reality to a science lab simulation causes more presence but less learning. Learn Instr 60:225–236. https://doi.org/10.1016/j.learninstruc.2017.12.007

Matusiak KK (2013) Image and multimedia resources in an academic environment: a qualitative study of students' experiences and literacy practices. J Am Soc Inf Sci Technol 64(8):1577–1589. https://doi.org/10.1002/asi.22870

Miltiadou M, Savenye WC (2003) Applying social cognitive constructs of motivation to enhance student success in online distance education. AACE J 11(1):78–95

Moore MG (1989) Three types of interaction. Am J Dist Educ 3(2):1–6

Moreno R, Mayer R (2007) Interactive multimodal learning environments. Educ Psychol Rev 19(3):309–236

Nehm RH, Ha M, Mayfield E (2012) Transforming biology assessment with machine learning: automated scoring of written evolutionary explanations. J Sci Educ Technol 21(1):183–196

Nouri J (2018) Students multimodal literacy and design of learning during self-studies in higher education. Technol Knowl Learn 24:683. https://doi.org/10.1007/s10758-018-9360-5

Ogor EN (2007, September) Student academic performance monitoring and evaluation using data mining techniques. In: Electronics, robotics and automotive mechanics conference (CERMA 2007), IEEE, pp 354–359

Olasoji R, Henderson-Begg S (2010) Summative assessment in second life: a case study. J Virtual Worlds Res 3(3). https://doi.org/10.4101/jvwr.v3i3.1460. https://journals.tdl.org/jvwr/index.php/jvwr/article/view/1460/1783

Olson S, Riordan DG (2012) Engage to excel: producing one million additional college graduates with degrees in science, technology, engineering, and mathematics. Report to the President. Executive Office of the President. https://files.eric.ed.gov/fulltext/ED541511.pdf

Outlaw J, Duckles B (2017) Why women don't like social virtual reality: a study of safety, usability, and self-expression in social VR. The Extended Mind, Portland. https://extendedmind.io/social-vr

Phuong AE, Nguyen J, Marie D (2017) Evaluating an adaptive equity-oriented pedagogy: a study of its impacts in higher education. J Effect Teach 17(2):5–44

Pintrich PR, De Groot EV (1990) Motivational and self-regulated learning components of classroom academic performance. J Educ Psychol 82(1):33–40

Porter ME, Heppelmann JE (2017, November 1) Why every organization needs an augmented reality strategy. Harv Bus Rev. https://hbr.org/2017/11/a-managers-guide-to-augmented-reality

Prain V, Cox P, Deed C, Dorman J, Edwards D, Farrelly C, Keeffe M, Lovejoy V, Mow L, Sellings P, Waldrip B, Yager Z (2013) Personalised learning: lessons to be learnt. Br Educ Res J 39:654–676

Radu I (2014) Augmented reality in education: a meta-review and cross-media analysis. Pers Ubiquit Comput 18(6):1533–1543

Rose H (1995) Assessing learning in VR: towards developing a paradigm virtual reality roving vehicles (VRRV) project. Human Interface Laboratory

Rosen JA, Glennie EJ, Dalton BW, Lennon JM, Bozick RN (2010) Noncognitive skills in the classroom: new perspectives on educational research. RTI International. PO Box 12194, Research Triangle Park, NC 27709-2194

Ryan A (2019, January 27) Thoughts on accessibility issues with VR. https://ablegamers.org/thoughts-on-accessibility-and-vr/

Sampson D, Karagiannidis C, Kinshuk (2002) Personalised learning: educational, technological and standardisation perspective. Interact Educ Multimedia 4:24–39

Sankey M, Birch D, Gardiner M (2010) Engaging students through multimodal learning environments: the journey continues. In: Proceedings of ASCILITE—Australian society for computers in learning in tertiary education annual conference 2010, pp 852–863

Shea P, Bidjerano T (2009) Community of inquiry as a theoretical framework to foster "epistemic engagement" and "cognitive presence" in online education. Comput Educ 52:543–553

Shermis MD, Burstein J, Higgins D, Zechner K (2010) Automated essay scoring: writing assessment and instruction. Int Encycl Educ 4(1):20–26

Spector C (2017, December 5) Students' early test scores don't predict academic growth over time. https://news.stanford.edu/2017/12/05/students-early-test-scores-not-predict-academic-growth-time/

U.S. Equal Employment Opportunity Commission (n.d.) The Rehabilitation Act of 1973. https://www.eeoc.gov/laws/statutes/rehab.cfm

U.S. General Services Administration (2018, November). IT Accessibility Laws and Policies. https://www.section508.gov/manage/laws-and-policies

Warburton S (2009) Second Life in higher education: assessing the potential for and the barriers to deploying virtual worlds in learning and teaching. Br J Educ Technol 40(3):414–426

Ward TB, Sonneborn MS (2009) Creative expression in virtual worlds: imitation, imagination, and individualized collaboration. Psychol Aesthet Creat Arts 3:211–221. https://doi.org/10.1037/a0016297

Weise K (2019) You're hired. Now wear this headset to learn the job—The New York Times [Newspaper]. https://www.nytimes.com/2019/07/10/business/microsoft-hololens-job-training.html. Accessed 11 July 2019

Winkelmann K, Keeney-Kennicutt W, Fowler D, Macik M (2017) Development, implementation, and assessment of general chemistry lab experiments performed in the virtual world of second life. J Chem Educ 94(7):849–858. https://doi.org/10.1021/acs.jchemed.6b00733

Yee N, Bailenson JN (2006) Walk a mile in digital shoes: the impact of embodied perspective-taking on the reduction of negative stereotyping in immersive virtual environments. In: Proceedings of PRESENCE 2006: the 9th annual international workshop on presence, Cleveland, 24–26 Aug

X-FILEs Jam: Ideation Process and Outcomes

Lawrence C. Ragan and Lorraine J. Ramirez Villarin

1 Motivations

In exploring the future of STEM education in 2026, the research team recognized the importance of gaining insights from a variety of stakeholders including administrators, faculty, instructional design and technology professionals, and visionaries who scan and identify trends and forces affecting higher education. Perhaps the most critical perspectives, however, were the insights, thoughts, opinions, and hopes of the user population, the students. The 2018 X-FILEs Workshop included a student-led discussion panel providing their perspective of STEM education in 2026. Four students, representing high school through graduate students, served as panelists and shared their hopes and expectations of the higher education experience. The X-FILEs research team recognized the value of this panel to the X-FILEs project and determined to structure an event that would stimulate and harness ideas directly from undergraduates and graduates. The target population was expanded from students representing STEM fields to students representing a range of backgrounds and academic disciplines.

The purpose of the student-based event was to gain complementary input that would enhance the data gathered at the 2018 X-FILEs Workshop. The form and structure of the 2018 Workshop would not be appropriate as it did not match the interest and logistics of a 1-day format. The project team settled on the "jam" model of ideation to organize and operate the event. Jams, sometimes referred to as "ideathons" or "designathons," focus on idea development rather than the produc-

L. C. Ragan (✉)
Ragan Education, State College, PA, USA
e-mail: lcr1@psu.edu

L. J. Ramirez Villarin
University of North Georgia, Dahlonega, GA, USA
e-mail: Lorraine.RamirezVillarin@ung.edu

© The Author(s) 2021
J. Ryoo, K. Winkelmann (eds.), *Innovative Learning Environments in STEM Higher Education*, SpringerBriefs in Statistics,
https://doi.org/10.1007/978-3-030-58948-6_7

Fig. 1 X-FILEs Student Panel Session

tion of a product. Jams are team-based, loosely structured exercises conducted in a face-to-face environment designed to bring out participants' creativity for developing innovative solutions to complex problems (Morrison 2009).

Jams engage students and promote creativity by giving them a challenge statement and structured guidance for solving it. The most creative jams require a multidisciplinary team in order to achieve the desired outcomes. Challenges require participants to apply knowledge to solve real-world problems. The X-FILEs Jam focused on a modified challenge statement derived from the 2018 X-FILEs Workshop (Fig. 1).

> "Using what you've learned about innovative learning environments, create a solution that improves or enhances the student experience for a challenging dimension of college-level STEM education."

2   Goals

Applying the challenge statement to the X-FILEs Jam program, the goal was to generate a collection of student idea-solutions where one or more of four technology categories may be applied to an aspect of teaching and learning to address a specific problem in STEM education. During the process, the research team gathered the participants' ideas and brainstorming results through a series of written tasks to track common patterns and keep a record of students' thoughts. Findings reflected what students envisioned of their STEM education experience in the near future.

3 Program Summary

3.1 Technology and Teaching and Learning Framework

In order to maintain connective themes with the original 2018 X-FILEs Workshop, the same four technology categories and the aspects of teaching and learning with their definitions were incorporated. Technology categories included personalized/adaptive learning, multimodal learning, cross reality, and artificial intelligence/machine learning. The aspects of teaching and learning encompassed content presentation, interactions and communications, learner activities, assessment, and co-curricular activities. The student teams were encouraged to address at what point in the course experience these technologies (individually or a creative mix) may be applied. The project's principal investigators, Drs. Ryoo and Winkelmann presented the definition and descriptions of these categories as they applied to the Jam. A video recording is available on the Jam website (Ryoo n.d.).

3.2 Design

Students, working in teams of three or four individuals, were guided through the day's ideation process in order to facilitate the successful completion of their idea-solution addressing the X-FILEs Jam challenge statement. Team idea-solutions were incubated and developed over a period of 6 to 8 h and presented at the conclusion of the Workshop. Award categories and prizes were described at the beginning of the program with the final awards being presented during the wrap-up activity.

3.3 Jam's Summary of Events

The X-FILEs Jam began with a tour and hands-on experiences guided by Martin Gallagher, director of the Evans Library Digital Scholarship Lab on the Florida Tech campus. He introduced the students to the four categories of emerging technologies discussed in the 2018 Workshop. This presentation was followed by participants exploring, interacting with, and reflecting on how today's technologies can become vital resources on future STEM higher education instruction.

Lawrence C. Ragan, the day's facilitator, introduced the X-FILEs research team and sponsors, followed by a review of the events, timing, and desired outcomes. The principal investigators, Drs. Ryoo and Winkelmann gave presentations to familiarize students with the transformative technologies in STEM education featured in the 2018 X-FILEs Workshop. Definitions and the potential of each technology, as well as examples, were provided.

The "STEM Idea Harvest" involved students in a gallery walk activity with stations representing each of the five aspects of teaching and learning. As the students

Fig. 2 Jam participants engaging in the gallery walk activity at the assessment station. February 8, 2020

arrived at each station, they reflected on the challenges, barriers, and opportunities presented by that aspect. They wrote their many ideas on sticky notes at each station. Students moved to all five stations in 3- to 5-min increments (Fig. 2).

In order to form the day's teams, participants were asked to select one of the aspects based on their personal interests. The desire was to form teams with a mix of discipline interests and levels of undergraduate or graduate education. Participants organized into five teams:

- Team Stuff representing content presentation
- Team WhatsApp representing interactions and communications
- Team To-Do representing learner activities
- Team Measure representing assessment remove evaluation
- Team Extra Extra representing co-curricular activities

Team members introduced themselves and developed a "How Might We ..." starting statement for their team-based on the generic example:

"How might we improve learning for students to enjoy, embrace, and apply STEM concepts using a variety of existing and future technologies?"

The day's agenda consisted of a series of activities designed to promote team formation, ideation, solution creation, and idea refinement. The beginning activity was the "Innovation Disruption Analysis." Teams randomly selected three cards from the Business Innovations Cards stack (Board of Innovation 2020). Each card provided an example of a case study from a company that generated a new product after recognizing a gap in the industry. The cards represented the company's responses to technological, market, customer, and regulation trends. Teams reflected on why it was an inspiring case and how it became an innovative disruption. They later presented a short description on how these disruptions may influence their idea-solution. The facilitators declared Team To-Do, focusing on the aspect of learner activities, winners after suggesting how the Duolingo app could be incorporated into each of the aspects of teaching and learning to improve the students' learning experience.

Participants continued with an "opposite thinking" exercise. Teams developed a list of assumptions for their aspect of teaching and learning. Based on these assumptions, the teams formulated the opposite response or representation of the assumption. For example, if one assumption of STEM education is that all students receive homework assignments using the same context, an opposite response may be that the context is personalized for each student based on their personal interest or area of study. Each team was provided writing materials and encouraged to record their ideas and the process as they developed their idea-solution plans.

To conceptualize the idea-solution, teams engaged in the "Amazing Idea-Solution Process." Inspired by the show *The Amazing Race*, the task involved teams independently completing certain steps to narrow their ideas before moving forward (Fig. 3). Five cards were created with instructions on how to complete each steps. The facili-

Fig. 3 Participants from Team WhatsApp working on Card 2 of the "Amazing Idea-Solution Process." February 8, 2020

tators provided the next level card to teams after step completion. These cards for the idea-solution development were:

Card 1. Solution identification
Card 2. Idea-solution/tech match
Card 3. Idea-solution user
Card 4. Amazing idea development
Card 5. Present with confidence!

At the completion of the "Amazing Idea-Solution Process," teams presented their idea-solutions to the larger community in a 10-min presentation and a 2-min question and answer session. While the participants voted for the Student's Choice Award, five judges evaluated the presentations using a criteria checklist to determine the winners in the remaining categories. Drs. Ryoo and Winkelmann presented the awards which would earn a gift card to each student in a team:

1. Best overall
2. Best technology
3. Most creative and innovative
4. Most impactful
5. Student choice

4   Outcomes

4.1   Trends Generated from the STEM Idea Harvest

The future of innovative learning environments within each aspect of teaching and learning presented challenges, barriers, or problems for students. Many of these were self-evident, such as the high cost of study materials or the lack of access to digital class notes. Others were more subtle and may only affect certain individuals or populations, such as students with physical or emotional disabilities and access to class materials while being deployed in the military. The following list represents a sampling of the recurring themes captured from the participants' notes.

Content Presentation Context Presented to Students: "When considering how the course content is managed and taught in a course, what issues, barriers, challenges, or problems do students have? For example, it may be the cost of materials, the timing of when you can get the materials or even the weight of the books! Include in this question the "teaching" of this course material. For example, too many lectures, not enough review time, lack of access to the professor, etc."

Common Themes of Participant Input:

- Time—Concerns were expressed around the impact of too little or too much time on a course. In general, when a set timeframe is established for all students, some will feel rushed through the materials, while others become impatient to move ahead. This perennial dilemma causes gaps in understanding or inhibits the ability to fully apply content knowledge within a real-life perspective.

- Cost—Participants identified the cost of a course and materials as a barrier to their educational experience. Some online programs may require additional subscriptions with fees not usually covered by the university.
- Resources—Participants also expressed a need for alternative sources of content, as textbooks quickly become outdated and obsolete. Additionally, course content may not always be available in multiple formats to accommodate students with disabilities or special needs.
- Relevancy—Students' most recurrent problem was the lack of relevance between course content and real-world tasks and applications.
- Additional suggestions included adapting the course content to the time allotted, consolidating material, personalizing or customizing content to students' needs, and making a more engaging or hands-on learning experience.

Interactions and Communications Context Presented to Students: "Thinking about the interactions and communications surrounding a course, what issues, barriers, challenges, or problems do students have? For example, not having easy access to the course schedule, not getting regular updates on course changes, not having access to other students' contact info, etc."

Common Themes of Participant Input:

- Time restrictions—There is limited access for individual meetings with instructors during office hours. One professor may not have time for all of his/her students.
- Communication anxiety or barrier—Students expressed concerns that communication and understanding may be affected when there is "contact apprehension" between teacher and student due to culture, language, power difference, or social anxiety.
- Lack of consistency—Other communication and interaction problems identified included the absence of consistency of style and format among educators. Different faculty use different methods, tools, and styles to interact with their students. This inconsistency of feedback from instructors makes it difficult for students to navigate course activities. Participants believe that learner and educator expectations are different and lack to portray real-world outcomes.
- Suggestions for improvement included the mandated use of Learning Management Systems, standardized versions of syllabi formats, and improved access to instructors.
- Participants also suggested providing increased customization of the student portal (e.g. individualized course interface).

Learner Activities Context Presented to Students: "The student activities include course assignments or requirements that encourage students to apply, practice and further develop facility with the course objectives. What issues, barriers, challenges, or problems do students face in this aspect of the course? For example, lack of time to complete assignments, no support available when needed, etc...."

Common Themes of Participant Input:

- Response Timeliness—Students expressed frustration with the lack of assistance and feedback when needed. Students related a desire for 24-hour help, especially for distance learners.
- Relevance—The second most common complaint was the absence of learner activities with real-life applications or relevance.
- Guidance/instructions—Suggestions involved designing clearer instructions and guidance on assignments.
- Engaging Activities—Students expressed a desire for more engaging activities that incorporate technology (e.g. VR, apps) and work related to the field of study.
- Suggestions included making learner activities more relatable to students' hobbies and backgrounds where possible. Another suggestion was for activities to be compliant with learner's pace and more collaborative work to reduce homework time.

Assessment Context Presented to Students: "Courses always include some method(s) of assessing progress in the course. Other methods may be used such as midterms or finals to determine the final grade. What issues, barriers, challenges, or problems do students have in regards to the assessment and evaluation methods used in a course?"

Common Themes of Participant Input:

- Testing Environment—Current evaluation and assessment strategies do not frequently take into account the uniqueness of each individual. Treating all students the same eliminates the ability to allow students with differences to show what they are capable of accomplishing.
- Fairness—Participants proposed adjusting grading on an as-needed basis by the instructor, for example, adding a curve scale when necessary or using equal weighting.
- Suggested alternative assessments comprised the use of tests based on real-world scenarios, making more personalized assessments based on student experiences and incorporating VR and other technology assessments to decrease costs and promote a more ethical curriculum (e.g., eliminating real-life animal dissections). Other suggestions included simple adjustments as modifying the testing environment, allowing additional time, and assessing skills rather than rote memory.

Co-curricular Activities Context Presented to Students: "Many times courses will include materials or experiences beyond the scope of the actual course. These may be related to the course content while other times they are "nice to haves" but not critical. For example, attending a lecture or presentation given by a visiting scholar or joining the BioResearch Club. What issues, barriers, challenges, or problems do students have with co-curricular activities?"

Common Themes of Participant Input:

- Limited Access/Lack of Resources—For some students, the distance to club meetings and activities (e.g., field trips) is not within reasonable limits. Virtual reality may allow remote participation, but not everybody has access to the equipment.
- Cost—Engaging in extracurricular activities can place an increased cost burden on the student that inhibits their participation. In some cases these activities are mandated (e.g., conferences, field trips) or highly recommended, and the student may not have the funds to participate.
- Scarce Industry Opportunities—identifying suitable opportunities to engage with intern programs is challenging with limited access to data. Students need flexibility and agility in finding a program that fits their academic and personal circumstances.
- Time—As students are increasingly occupied with work, family, and academic activities, finding the time to participate in extracurricular relationships presents a barrier.

4.2 Amazing Idea-Solutions

Five teams of three or four individuals were organized as an output of the aspects of teaching and learning Idea Harvest. Each team chose a single aspect to use as the context for their idea-solution. The participants' thoughts garnered on post-it notes through the gallery walk became the basis for the day's work. The remaining sequence of the day's events guided the teams through the ideation process in order to present a final idea-solution for the Jam competition.

Content Presentation Team Stuff designed a "Fitbit"-type device that could track a student's learning pace and could suggest "learning time" that an individual should devote to a course or task.

Interactions and Communications Team WhatsApp conceived an app that would connect students 24-hours a day with either live or AI tutors worldwide to provide assistance clarifying content or requiring help related to course materials.

Learner Activities Team To-Do addressed a modification to Learning Management Systems via a survey to customize student's learning activities to their hobbies, interests, or careers making them more meaningful and engaging. This idea was premised on the belief that students would be more interested and motivated with their learning when it directly interacted with their personal interests.

Assessment Team Measure's idea consisted of creating a "controlled virtual reality environment" for administering formative assessments to students requiring special accommodations (e.g. ambient music and relaxing atmosphere, tasks assessed virtually, track eye movement with Oculus to denote attention, etc.)

Co-curricular Activities Team Extra Extra formulated an app called "Mentor Tinder." This app would consist of a database capable of matching a student with appropriate mentors that shared their interests and expectations. These mentor relationships could be local, regional, or even international.

5 Recommendations

The X-FILEs Jam was considered a success in gathering input from the end users, the students. There are features of the event that, slightly modified, may have produced different outcomes. The research team makes the following recommendations seeking to improve the experience for students as well as more closely align with the original X-FILEs project goals.

1. The challenge statement used as the basis of the Jam focused the student team's efforts on designing an idea-solution to a specific problem in STEM education. A modification to the Jam challenge statement to more closely align with the statement used for the X-FILEs Workshop would be beneficial in producing complementary outcomes from the Jam experience.
2. Leading individuals through the ideation process requires time for reflection and deeper thinking. Recommended strategies to enable additional "incubation" time for teams to consider new or novel idea-solutions would include forming teams as an outcome of pre-Jam webinars and enabling discussion/forum boards to encourage interaction prior to the Jam event.
3. Predictably, without adequate encouragement and time, teams gravitated toward "top-of-mind" and obvious idea-solutions. Including a process to share and review early ideas with program organizers or other guests would guide teams toward more new or novel solutions.
4. Conduct pre-Jam events including an online design competition in order to stimulate creative thinking and jumpstart the idea-solution development at the event.
5. Establish at least two teams working on the same dimension of an aspect of teaching and learning. At the appropriate time, combine these teams into one in order to generate a new and novel idea-solution.
6. Require teams to conduct research or perform literature reviews to determine if the idea-solution may already exist.
7. Provide teams with access to the technologies they have employed in order to test the feasibility of in their idea-solution.
8. Place increased emphasis on the unique and innovative nature of the desired outcomes throughout the Jam events.
9. Increase the number of participants through a variety of strategies including introducing the team competition during the pre-Jam webinars (Fig. 4).

Fig. 4 X-FILEs Jam Organizers and Participants

6 Conclusion

Gaining the insights and input of the end user of any system is a critical step in product design. Soliciting the student voice in describing and envisioning the future of STEM education through a Jam event proved to be an effective and educational experience for all. The success of the program was largely due to the willingness of the participants to address the challenge by generating creative and practical idea-solutions. Although the number of student participants was less than antici-pated from preregistration data, they dedicated the day to working as a team to consider how to improve the state of STEM education through the use of the four technology categories. The quality of their presentations of idea-solutions validated the process and the efforts of each team.

Capturing the student output through the transcription of notes and drawings provided a glance into the ideation progress and outcomes from each team. The positive interactions among participants and their willingness to consult with the X-FILEs research team served as a valuable guiding force throughout the event. Final evaluations denoted a high sense of satisfaction by students, suggesting the participants enjoyed and found value in the day's activities. Evaluations encom-passed everything from the accommodations to the facilitators and the overall view of the event.

As in every event offering, reflection on the areas for improvement presents an opportunity to enhance and enrich the event design. The recommendations suggest minor adjustments to a Jam event in order to more closely align the designed activities to the larger program goals.

References

Board of Innovation (2020) Brainstorm cards. https://www.boardofinnovation.com/tools/brainstorm-cards

Morrison K (2009) The implications of 'jam' and other ideation technologies for organisational decision making. Cult Sci Journal 2(1):1. https://doi.org/10.5334/csci.21

Ryoo J (n.d.) eXploring the Future of Innovative Learning Environments (X-FILEs) Jam—Student-generated solutions to challenges in STEM higher education. https://sites.psu.edu/filejam/. Accessed 6 Apr 2020

Index